Dialectical Behavior Therapy

What You Need to Know About DBT and a Simple Guide to Cognitive Behavioral Therapy

Contents

PART 1: DIALECTICAL BEHAVIOR THERAPY 1

INTRODUCTION 2

CHAPTER ONE: DIALECTICAL BEHAVIOR THERAPY AND MINDFULNESS FOR HANDLING EMOTIONS 4

CHAPTER TWO: EMOTIONAL AND MENTAL HEALTH DISORDERS-SIGNS TO LOOK OUT FOR 11

CHAPTER THREE: SETTING GOALS FOR EMOTIONAL AND MENTAL WELLNESS 21

CHAPTER FOUR: ANXIETY DISORDERS – 8 DBT TECHNIQUES FOR INSTANT RELIEF 27

CHAPTER FIVE: DEPRESSION AND EMOTIONAL REGULATION – 7 DBT TIPS TO FEEL BETTER NOW 33

CHAPTER SIX: WORKPLACE STRESS: 9 WAYS TO USE DBT AT WORK 40

CHAPTER SEVEN: BORDERLINE PERSONALITY DISORDER: TAME IMPULSIONS AND MOOD SWINGS WITH DBT 46

CHAPTER EIGHT: IMPROVING YOUR DBT DISTRESS TOLERANCE SKILLS 53

CHAPTER NINE: MINDFULNESS TOOLS FOR FEAR, INSECURITIES, AND PHOBIAS 59

CHAPTER TEN: MINDFULNESS MEDITATION TECHNIQUES FOR ANXIOUS MINDS 66

CHAPTER ELEVEN: OCD – 11 MINDFUL WAYS TO BEAT THE OBSESSIVE MIND 72

CHAPTER TWELVE: HOW TO STOP A PANIC ATTACK WITH MINDFULNESS 78

CHAPTER THIRTEEN: TRAUMA AND PTSD – HOW DBT AND MINDFULNESS CAN HELP 84

CHAPTER FOURTEEN: RELAPSE PREVENTION 91

CONCLUSION 98

RESOURCES 100

PART 2: COGNITIVE BEHAVIORAL THERAPY 102

INTRODUCTION 103

CHAPTER ONE: WHY USE CBT? 106

CHAPTER TWO: IDENTIFYING MENTAL HEALTH DISORDERS 114

CHAPTER THREE: GOAL SETTING: YOUR STARTING POINT TO MENTAL AND EMOTIONAL WELLNESS 123

CHAPTER FOUR: ANXIETY AND WORRY: CBT TECHNIQUES TO REDUCE BOTH NOW 131

CHAPTER FIVE: DEALING WITH DEPRESSION: CBT TIPS TO FEEL BETTER INSTANTLY 140

CHAPTER SIX: WORKPLACE CBT: WAYS TO BEAT STRESS AT WORK 149

CHAPTER SEVEN: INTRUSIVE THOUGHTS: ACKNOWLEDGING AND ELIMINATING THEM WITH CBT 158

CHAPTER EIGHT: MINDFULNESS AND THE CBT CONNECTION ... 167

CHAPTER NINE: THREE MINDFULNESS MEDITATION TECHNIQUES YOU SHOULD KNOW 175

CHAPTER TEN: DON'T PANIC! HOW TO STOP A PANIC ATTACK WITH MINDFULNESS 186

CHAPTER ELEVEN: HOW TO PREVENT A RELAPSE 196

CONCLUSION 203

REFERENCES 204

Part 1: Dialectical Behavior Therapy

An Essential DBT Guide for Managing Intense Emotions, Anxiety, Mood Swings, and Borderline Personality Disorder, along with Mindfulness Techniques to Reduce Stress

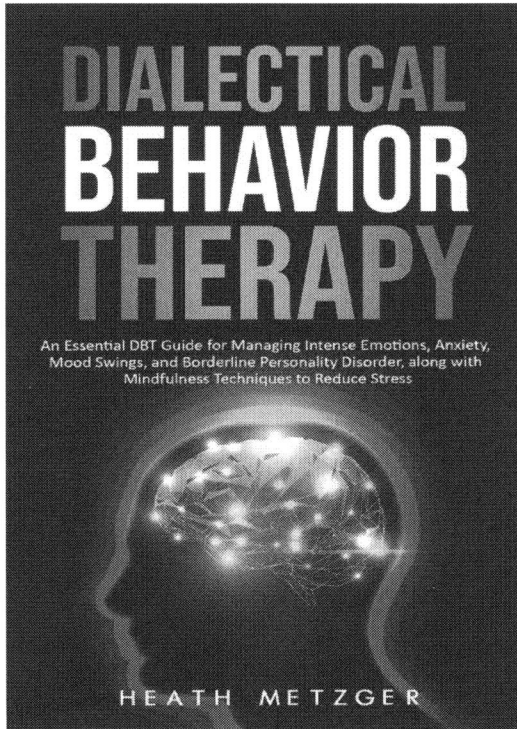

DIALECTICAL BEHAVIOR THERAPY

An Essential DBT Guide for Managing Intense Emotions, Anxiety, Mood Swings, and Borderline Personality Disorder, along with Mindfulness Techniques to Reduce Stress

HEATH METZGER

Introduction

Take a moment and think of a life without any anger, disappointment, stress, distress, frustration, or any other undesirable emotions. Think about how wonderful your life would be if you could control your emotions. Even if they seem uncontrollable right now, you can learn to regulate them. Living a life like this certainly sounds great, doesn't it? If you want to do this, then this is the perfect book for you.

There are a variety of emotions we all experience; some of them are desirable, while others are undesirable. Emotions tend to directly or indirectly influence different aspects of our lives. Since our thoughts are often based on our emotions, it can become difficult to stay rational when your emotions are running high. The inability to cope with intense emotions can quickly hamper your ability to lead a happy and stress-free life.

Millions of people across the globe suffer from a variety of emotion regulation disorders like borderline personality disorder (BPD), obsessive-compulsive disorder (OCD), post-traumatic stress disorder (PTSD), anxiety, and depression. If you are tired of allowing your emotions to guide your decisions and want to learn to control them, the DBT method will come in handy. DBT stands for Dialectical Behavior Therapy. Dr. Marsha Linehan developed the concept of DBT, and it is a clinically proven and evidence-based treatment for emotion regulation and managing intense emotions.

This book is ideal for anyone who wants to learn more about DBT. The information included in this book is presented in an easy to understand fashion, making it ideal for beginners and experts. This book is your go-to guide about DBT and mindfulness. The up-to-date information, along with different techniques—including mindfulness—to deal with various mental disorders, makes it different from other guides available on the market. Mindfulness is not just integral to DBT; it is vital for your overall wellbeing. Mindfulness teaches you to live your life in the present, instead of dwelling on worrisome thoughts about your past or the future. If you cannot live your life in the present, then your thought patterns will be riddled with negativity and anxiety.

In this book, you will learn about the history of DBT, the meaning and core principles of DBT, the benefits it offers, different types of mental health issues, and how DBT can be used to tackle and manage them. The awareness about mental health is steadily increasing, and with it, new disorders are being identified. Long gone are the days when mental health issues were viewed as a taboo. From dealing with BPD to tackling OCD and anxiety, to managing PTSD, insecurities, phobias, and depression, this book has everything you have been looking for. All the practical tips and techniques given in this book focus on mindfulness and DBT. All that's required is a serious commitment, as well as time, effort, patience, and consistency, to handle any difficult or unpleasant emotions.

If you are eager to take charge of your life and handle your emotions without allowing them to overwhelm you, then let us get started immediately.

Chapter One: Dialectical Behavior Therapy and Mindfulness for Handling Emotions

Dialectical Behavior Therapy (DBT) is a form of cognitive-behavioral therapy that was originally invented to assist those individuals who have borderline personality disorder (BPD). The primary goal of DBT is to teach an individual to live in the moment, cope with stress, regulate emotions, and improve relations with their own self and others. Even though it was originally intended for those with BPD, these days, it is used for any other health conditions where an individual exhibit any form of self-destructive behavior or substance abuse. So, it is safe to say that DBT can help anyone manage and handle their emotions constructively.

The premise behind this approach is that some individuals are inclined to react intensely and disproportionately toward certain situations, especially the emotional ones associated with different relationships in their lives. The theory of DBT suggests that certain individuals have more rapid arousal levels toward situations than an average individual. Therefore, they experience a higher level of emotional stimulation and take a while longer than an average individual to return to the normal state.

History

Dr. Marsha Linehan and her colleagues introduced the concept of DBT during the 1980s. They discovered that certain aspects of cognitive-behavioral therapy (CBT) were not sufficient when dealing with patients diagnosed with BPD. So, Dr. Linehan, along with her team, came up with various techniques and a new course of treatment to help meet the unique needs of those with BPD. The primary concept of this technique is based on philosophical processes known as "dialectics." Dialectics essentially suggest that all things are made of opposites, and change occurs whenever an opposing force is stronger than its opposite force. In academic terms, it can be described as thesis, antithesis, and synthesis. There are three basic assumptions on which dialectics are based, and they are as follows:

- Everything is interconnected.
- Change is not only constant but it is inevitable.
- All opposites can be integrated to form a close approximation of the truth.

In DBT, the patient, along with the therapist, actively works on resolving any contradictions between self-acceptance and change, to bring about a positive change in the patient. Dr. Linehan and colleagues also came up with another technique known as "validation." They observed that when the need for change was coupled with validation, a patient's cooperation increased, and any distress associated with coping with change was reduced.

Components of DBT

There are three components of DBT, and they are as follows.

DBT is support oriented. It essentially enables an individual to identify their strengths and develop them so that they not only feel better about themselves, but their life in general.

DBT is cognitive-based. It helps an individual to learn skills to identify their thoughts, beliefs, or any other assumptions they have, which make their life difficult. For instance, a thought like, "I'm a terrible person if I feel angry," or "I need to be good at everything, and anything short of perfection is undesirable," can immediately make anyone feel terrible about themselves or their lives. DBT enables a person to identify such patterns of thinking and replace them with different patterns of thinking that will make life bearable. For instance, the negative thought patterns discussed above can be replaced with something more constructive like: "I don't need to be perfect for others to like me," or "Anger is a normal emotion that everyone experiences."

There is plenty of collaboration required in DBT. Its success essentially depends on the relationship between the client and the therapist. In DBT, individuals are encouraged to figure out the problems in their relationships with the help of a therapist. In DBT, an individual need to complete their homework, try different activities suggested by the therapist, and practice certain self-soothing skills when upset. All these aspects are crucial to DBT, and they are taught in weekly lectures or sessions followed by a review. DBT is a great way to rewire one's brain and replace harmful thought patterns with positive ones. That, in turn, will have an overall positive effect on the life of an individual.

How Does DBT Work?

DBT has now finally evolved into a regular type of cognitive-behavioral therapy. A regular course in DBT takes around 24 weeks, but there are varying lengths for the treatment. It is a careful blend of individual and group sessions. Whenever an individual opts for DBT, they will be expected to participate in the three therapeutic settings, as mentioned below.

A classroom setting is where an individual is taught certain behavioral skills via homework assignments, along with different role-playing ways to interact with others. Usually, it lasts anywhere between two to three hours weekly.

In individual therapy, a trained professional or therapist uses the behavioral skills taught in the previous sessions to help the person overcome any personal challenges in life. These sessions run concurrently with the classroom work. A usual therapy session lasts for up to 60 minutes and is conducted once a week.

The third option is on-call coaching. The on-call approach allows an individual to seek their therapist in between any of the sessions to receive guidance to cope with any difficulties they face in the moment.

DBT is not only helpful for the patient but the therapist too. It often offers support to therapists as they navigate any complicated issues; they can meet with a consultation team to sustain their motivational levels while treating their patients.

Modules of DBT

Interpersonal Effectiveness

If you wish to become more assertive in any relationship in your life, then interpersonal effectiveness is essential. It is about dealing with yourself, respecting yourself, and those around you. It enables you to understand your boundaries while maintaining healthy and positive relationships. This occurs once you start listening and communicating effectively and efficiently.

Here's a simple exercise you can try. If you want to improve your relationships via positive communication, then use the acronym GIVE. GIVE stands for gentle, interest, validation, and easy. While communicating with others, be gentle and never judge, attack, or threaten the other person. Always show interest via good listening skills. It could be something as simple as listening without interrupting while the other person is talking. The third thing to keep in mind is validation. Regardless of whether you agree or disagree with the other person's feelings or thoughts, always acknowledge them. The final thing you must keep in mind is to go easy. Always have an easy attitude in life, smile more often, and don't take things too seriously.

Mindfulness

Mindfulness is perhaps the most important principle of DBT. It enables you to focus only on the present and start living your life in the moment. By doing this, you can easily notice all the thoughts you think, the feelings or sensations you experience, your impulses, and the world around you. Mindfulness essentially enables you to calm your mind and come up with healthy coping mechanisms to deal with emotional chaos or pain. It also enables you to stay calm and avoid engaging in any negative thought patterns or impulsive behaviors.

Here is a simple exercise you can try. To develop mindfulness, start concentrating on your breathing. Observe the way you feel whenever you inhale and exhale. Notice the way your belly rises and falls as you breathe in and breathe out. By shifting all your attention to your breath, it enables you to stay grounded in the moment, while letting go of unnecessary thoughts.

Emotion Regulation

As the name suggests, emotion regulation is all about understanding and regulating your emotions. Unless you do this, you cannot maintain your emotional wellbeing. It enables you to adjust your emotions, along with their intensity, and to regulate your responses. By observing, and coping with negative emotions, you can increase the likelihood of positive emotional experiences, while reducing unnecessary emotional vulnerability.

Here's a simple exercise you can try. Take a moment and notice how you are feeling, and think of the opposite of whatever you are feeling. If you feel sad and want to withdraw from your usual circle of friends, try doing the opposite. Instead of withdrawing, make plans to meet your loved ones.

Distress Tolerance

A common issue a lot of people face is accepting oneself, and all the current situations in life. Distress tolerance will teach you to tolerate or overcome any crisis using simple techniques like self-soothing, distraction, adding movement, or thinking about the pros and cons. Distress tolerance will give you the skills required to cope with distressing or intense emotions, while maintaining a positive outlook on life.

Here's a simple exercise you can try. To improve your distress tolerance skills, try putting your body in charge for a change. If you are sitting indoors, then go outside for a while. If you are sitting at the desk, go for a short walk. You can also try running up and down a flight of stairs. You are essentially trying to distract your mind by permitting your emotions to flow freely through your body.

Benefits of DBT

Perhaps the most significant benefit of DBT is that it helps to get a clearer understanding of emotions, and the ability to pause and check on one's own emotions. If you can accept your reality and understand things the way they are, without resorting to intense reactions, then any inclination toward destructive behaviors will reduce. Apart from this, here are some of the other benefits of DBT.

DBT enables an individual to become less judgmental. Once you stop being judgmental and take on a neutral stance about yourself and the world around you, you will be better equipped to regulate emotions. Assuming a nonjudgmental stance makes you less susceptible to be ruled by your emotions. It is also believed that DBT can help curb or significantly reduce any suicidal thoughts or other destructive behaviors.

Once you have successfully completed the entire DBT program, you will be in a better position to form and nurture lasting relationships. Maintaining strong and consistent relationships certainly adds to one's mental health.

DBT also enables you to develop a healthy self-image. Once you start feeling better about yourself, the urge to indulge in substance abuse reduces. Apart from that, you'll come up with healthier coping mechanisms, instead of destructive ones, to deal with the happenings of day-to-day life.

Instead of letting go of your emotions, you will successfully learn to manage them. It also teaches you techniques to become more assertive in a constructive manner and come up with simple techniques for managing any personal conflicts. Self-acceptance is also a benefit offered by DBT. For instance, once you accept that you have certain flaws, and having these flaws doesn't make you a bad person, you will feel better.

DBT and Mindfulness

Mindfulness is all about paying attention to whatever is happening in the moment. It is about living your life in the present, without dwelling too much on your past or worrying about the future. Most of us tend to go through life on autopilot. While using mindfulness, you will shift from this autopilot mode, and learn to savor every moment of life. For instance, when you live your life on autopilot, you can travel from one destination to another and not remember a single part of the journey. With mindfulness, you start becoming aware of not just your thoughts, but also all your actions.

There might be instances where you feel a little overwhelmed by your emotions. Mindfulness enables you to step back from your feelings and carefully analyze the situation at hand. Once you notice whatever is happening, you can easily avoid allowing your emotions to get out of hand. You not only notice whatever is going on inside you, but the world outside of you as well. You will have a better idea of your thoughts, sensations, feelings, and impulses, and use all your senses while living your life. Some of the benefits offered by mindfulness include: reducing distractions; increasing emotional regulation; reducing unnecessary thoughts; dealing with emotions and anxiety or depression; reducing undesirable emotions; and increasing activity in the brain associated with positive emotion.

So, what does mindfulness have to do with DBT? As stated earlier, mindfulness forms the backbone of dialectical behavior therapy. In fact, it is the first skill that's usually taught in DBT. Without mindfulness, it will be impossible to change any long-standing patterns of thinking, feeling, and behaviors. It also enables you to regulate your emotions and get through any difficult situations in life, without making things worse for yourself. It is also essential for regulating any interpersonal conflicts.

Chapter Two: Emotional and Mental Health Disorders- Signs to Look Out For

Whenever we talk about health, we tend to think of physical or mental ailments. However, there is a third aspect of health, and it includes emotional health. Not a lot of people realize the importance of emotional health. In fact, the terms of mental health and emotional health are often used synonymously. Well, there is a difference between these two. Certain areas of mental health and emotional health are overlapping, and they share certain similarities, but these two concepts are quite different. Your overall wellbeing depends on the balance between your emotional, mental, and physical health.

Two important aspects of your personality that trickle into mental health, are your ability to process and reason. You require a strong sense of reasoning to understand your emotions, regulate them, and prevent instability. All the decisions you make about how you wish to react to different scenarios must be carefully processed, to avoid unnecessary stress and anxiety. Any imbalance tends to put your health in more than a precarious state, which influences your overall ability to function optimally. So, what is mental health?

Your cognitive thinking skills and the ability to stay focused are associated with your mental health. These skills also involve your ability to store information, process it, and understand all the information you absorb. Yes, psychological, emotional, and social wellbeing are all aspects of your mental health.

At times, people might experience certain issues associated with mental health, which negatively influences their thinking, mood, and overall behavior. It is believed that one out of five adults in the United States suffers from some form of mental disorder every year. There are different types of mental disorders, like schizophrenia, depression, bipolar disorder, and so on. You will learn more about these in the next section. Some of the most common symptoms of instability in mental health are acute mood swings, abuse of alcohol, use of drugs, hallucinations, harmful thoughts, suicidal tendencies, withdrawal from society, lower energy, feelings of hopelessness, excessive sleep, lack of sleep, and the inability to perform usual tasks effectively.

So, what is emotional health? Emotional health involves psychological functioning. It is about understanding oneself, emotions, and expressing them in an age-appropriate manner. Your behavior, thoughts, and feelings, both internal and external, are included in your emotional health. Managing your emotions, gauging your reactions, and preventing unnecessary stress, are all essential for managing your emotional health. The process of maintaining your emotional health is an ongoing one. If you want to lead a healthy and happy life, then you cannot afford to overlook your emotional wellbeing.

The concepts of emotional and mental health are distinct from one another, but unless there is cohesion between them, you cannot be balanced. By effectively managing your emotional and mental health, you can minimize anger, stress, anxiety, fear, worry, or any other undesirable emotions. We all tend to experience several hundreds of emotions and thoughts daily. Most of the decisions we make are based on these emotions and thoughts. They stem from our ability to cognitively reason and process all the information we receive from specific situations. Therefore, it is essential that you understand the different aspects of your life, to improve yourself. These two complementary, yet separate, aspects of your health work together to ensure your overall wellbeing. Apart from this, they influence the way you interact and communicate with yourself and others around you.

Dialectical behavior therapy provides certain techniques and tactics you can use to improve your emotional and mental health. That, in turn, improves your emotional quotient while enhancing the overall quality of your life.

Types of Emotional and Mental Health Disorders

There are various types of mental health and emotional health disorders. These days, people's awareness about the importance of emotional and mental wellbeing is steadily increasing. Long gone are the days when issues associated with mental health were viewed as taboo. All the ongoing research and development in this field is aimed at improving the overall quality of life of an individual. Learning about the following emotional and mental health disorders will put you in a better position to gauge your emotional and mental wellbeing.

Mood Disorders

Mood disorders are also known as affective disorders. They involve the feelings of constant and extreme levels of sadness or happiness and extreme fluctuations in one's overall mood. Bipolar disorder, cyclothymic disorder, and depression are common types of mood disorders. A psychiatrist or trained health professional can diagnose these types of affective disorders.

Anxiety Disorders

Stress, fear, or dread are the common reactions that people with anxiety disorders have toward specific situations or objects. Apart from this, they can also experience physical symptoms of panic or anxiety, such as excessive sweating and elevated heart rate. There are three straightforward ways to diagnose an anxiety disorder. The first is when an individual's response to a situation is inappropriate or exaggerated, the second is when an individual cannot control his response to a situation, and the third is when anxiety starts interfering with an individual's ability to function normally. There are different types of anxiety disorders, like social anxiety disorder, generalized anxiety disorder, specific phobias, and panic disorder.

Eating Disorders

Any extreme attitudes, behaviors, or emotions associated with food and one's body weight are associated with eating disorders. The most common types of eating disorders include bulimia, anorexia, and binge eating disorder.

Psychotic Disorders

Distorted patterns of thinking and awareness are the primary characteristics of a psychotic disorder. The most common symptoms of a psychotic disorder are delusions and hallucinations. In hallucinations, an individual tends to experience sounds or images that are not real, and they can even start hearing voices. A person may also have certain fixed beliefs that are proven to be false, but the person seems to accept them as the absolute truth, even if there is contrary evidence. The most common example of a psychotic disorder is schizophrenia.

Personality Disorders

An individual with a personality disorder tends to have inflexible and extreme characteristics that are not just distressing to the individual in question, but those around them as well. Personality disorders often interfere with one's ability to function optimally and effectively in different relationships in their personal and professional lives. Apart from this, an individual's thinking patterns, along with behaviors, tend to differ from the usual expectations of society. These patterns are often so rigid that they effectively interfere with an individual's ability to function normally. The most common examples of personality disorders include obsessive-compulsive personality disorder, antisocial personality disorder, and paranoid personality disorder.

Impulse Control Disorders

The inability to regulate and resist impulses or an urge to perform actions that can be potentially dangerous to yourself or others is known as an impulse control disorder. The most common examples of impulse control disorders include kleptomania, pyromania, and compulsive gambling. Drugs and alcohol are common sources of addictions. Usually, individuals with impulse control or addiction disorders get extremely involved with their addiction and start ignoring other aspects of their life regardless of the consequences.

Post-Traumatic Stress Disorder

Whenever an individual experience a terrifying or traumatic event, it can result in post-traumatic stress disorder or PTSD. The event could be physical assault, loss of a loved one, natural disaster, or a sexual assault, for example. Those who have PTSD, become emotionally numb because of the frightening thoughts or lasting damage associated with the memories of a traumatic event.

Obsessive-Compulsive Disorder

When an individual is constantly plagued by fears or thoughts that instigate them to perform specific rituals or routines consistently, it is known as an obsessive-compulsive disorder. Obsessive-compulsive disorder or OCD is the creation of certain obsessions or rituals known as compulsions. For instance, an individual might be plagued with an unreasonable fear of germs and, therefore, constantly washes or sanitizes their hands.

Dissociative Disorder

A dissociative disorder is primarily characterized by any severe changes in memory, memory disturbances, or any other disturbances in one's consciousness, identity, and overall awareness of oneself and their surroundings. Any event that causes overwhelming stress can result in a dissociative disorder. It is often the result of a traumatic accident, event, or disaster an individual either witnessed or experienced. It was formerly known as split personality disorder or multiple personality disorder. Depersonalization disorder is an example of a dissociative disorder.

Stress Response Syndrome

Stress response syndrome was formerly known as an adjustment disorder. Whenever an individual develops certain behavioral or emotional symptoms in response to any stress-inducing event or situation, it results in stress response syndrome. Stressors can be natural disasters, unfortunate events, major crises, or even interpersonal problems. Dealing with the death of a loved one, overcoming a natural disaster, dealing with a diagnosis of a life-threatening illness, or divorce can result in stress response syndrome.

It usually occurs within three months of the occurrence of a stressful situation or event, and is automatically terminated within six months or whenever the stress is eliminated or stopped.

Somatic Symptom Disorders

Somatic symptom disorder was formerly referred to as somatoform disorder or psychosomatic disorder. In this disorder, an individual tends to experience physical symptoms of pain or illness, to an extent that is excessive and disproportionate, but medical tests cannot find the physical cause. It induces great levels of distress and can make anyone miserable.

Factitious Disorders

In a factitious disorder, an individual might intentionally or knowingly complain of either emotional or physical symptoms to place himself in the role of a patient or someone who requires help and attention.

Sexual Disorders

Sexual disorders adversely affect an individual's sexual performance, behavior, and desire. Paraphilia and sexual dysfunction are the most common examples of sexual disorders.

Tic Disorders

The most obvious symptom of a tic disorder is making or displaying any sounds or non-purposeful body movements that are sudden, repeated, uncontrollable, and quick. Any sounds made involuntarily are known as a vocal tic. A common example of a tic disorder is Tourette's syndrome.

Common Causes

The human brain is extremely complicated and incredibly powerful. It is not easy to understand the exact causes of mental health issues. In fact, researchers are still trying to figure out the primary causes. The cause might not always be singular, and it is often a mix of different factors, including genetics, environment, childhood, or the way your brain works, and so on. Here are some of the most common factors that can result in mental health issues.

Genetic factors play an important role. If there is any history of mental health issues in your family, then it increases your risk of developing the same. However, just because someone in the family has a mental illness, doesn't mean others will as well. Certain medical conditions, along with hormonal changes, can also influence one's mental health. Substance abuse like excessive drinking or indulging in illicit drug use can trigger episodes of psychosis or manic episodes. Paranoia can also be induced by the consumption of drugs like amphetamines, marijuana, and cocaine. Any trauma or stress you might have dealt with in childhood or adulthood can also increase your risk of mental health issues. Traumatic experiences like serving in a war zone, or dealing with domestic violence, abusive relationships, or abuse in childhood can leave lasting scars on your psyche. Apart from that, there are certain personality traits, like a desire for perfection or low self-esteem that can increase the risk of anxiety or depression.

Understanding these causes is important, since it enables you to determine whether you are at risk of developing any mental health issues. Unless you understand and accept yourself and your reality, you cannot use the DBT techniques discussed in the subsequent chapters.

Common Symptoms

Now that you're aware of the different types of disorders, how do you identify them? How do you spot a difference between a bad mood and something potentially more serious? If you don't feel like yourself or maybe one of your loved ones doesn't seem like their usual self, what can you do? Indulging in excessive drinking, prolonged periods of sadness, social withdrawal, or indulging in negative thoughts and behaviors are all signs of some type of mental health issue. Mental health requires a serious commitment, and overlooking it is not prudent.

Feeling Depressed or Unhappy

One of the most common signs of any mental health issue is feeling unhappy or depressed for prolonged periods. We all tend to feel a little sad from time to time, and it could be for various reasons. However, holding onto such undesirable emotions for prolonged periods is often a symptom of mental health issues. If you notice that you don't feel like doing something you used to or feel inexplicably sad, don't ignore these symptoms. Perhaps you seem sad or irritable for a couple of weeks and don't have the motivation to go about your daily life. If you are running low on energy or get teary-eyed all the time, it might be a symptom of depression.

Feeling Worried or Anxious

Experiencing stress and worry is quite normal, and everyone experiences it at some point or another. However, prolonged anxiety is a sign of mental health issues. If your anxiety constantly interferes with your life or prevents you from leading a normal life, it is a symptom that requires your immediate attention. Other symptoms you can watch out for include constant headaches, diarrhea, heart palpitations, shortness of breath, and restlessness.

Problems with Sleep

An average adult requires about 7 to 9 hours of good quality and undisturbed sleep daily. If you notice any persisting changes in your usual sleep patterns, it can be a symptom of mental health issues. Insomnia is a common symptom of anxiety issues or even substance abuse. The inability to sleep, or sleeping excessively, are both undesirable. It could also be a symptom of a sleeping disorder or, in extreme cases, depression. As with any other function, there needs to be balance.

Emotional Outbursts

Our mood keeps changing, and everyone has different moods. However, any dramatic or sudden changes in mood like extreme anger or even distress are symptoms of mental health issues. Any extreme oscillations in emotions or inability to regulate your emotions is a warning sign. If you feel like your emotional responses to situations are rather severe and disproportionate to the issue at hand, it is time to evaluate your mental health.

Substance Abuse

There is nothing wrong with drinking occasionally or socially. However, there is a limit to every function, and if you exceed this limit, it will be harmful to your overall wellbeing. If you notice that you are using alcohol or drugs as a coping mechanism to deal with any issues in your life, it is a sign of trouble. Substance abuse is among the most common forms of undesirable coping mechanisms an individual can opt for due to any emotional or mental imbalances.

Urge to Withdraw

Spending some time by yourself is essential for recharging your energy. We all require a little alone time away from distractions. Being quiet occasionally is not worrisome. However, if you feel like you are withdrawing from life in general, and it isn't a normal change, it could indicate a mental health issue. If you are regularly isolating yourself or refuse to join any social activities, it could be a sign that you need some help.

Drastic Changes in Appetite or Weight

Most of us usually try to lose a couple of extra pounds, but extreme fluctuations in weight or rapid weight loss can be a potential warning sign for any form of mental health issue. Any drastic changes in appetite or weight loss are usually associated with eating disorders or depression. If you notice that you are using food as an unhealthy coping mechanism, it is time to evaluate your mental health.

Drastic Changes in Feelings or Behaviors

Usually, most of the mental health issues start as minute changes in the way you think, feel, and behave. If you notice any significant and progressive changes in the way you behave or if you feel you are behaving in a manner that isn't normal for you, you might be developing a mental health issue. If something doesn't feel right or you feel like you are missing out on something, it's time to seek professional advice and have a conversation about your mental health.

Feeling Guilty

You might feel guilty whenever you do something wrong. However, if the feelings of guilt or worthlessness don't go away, this could be a possible sign of mental health issues. For instance, if you notice you are entertaining thoughts like, "Everything is my fault," "I cannot be successful, I am a failure," or "I am worthless," for prolonged periods, it can indicate a deeper issue. If you feel like you are constantly blaming yourself and critically analyzing everything you do, it could be a sign of depression as well.

It is essential that you pay some attention to your mental and emotional wellbeing. Make a note of it whenever you experience any of the symptoms discussed in this section. Don't try to suppress the symptoms, and instead, seek professional advice and try to deal with them.

When to Seek Help

If the feelings of guilt and worthlessness are left unchecked, they can develop into suicidal thoughts. It can also prompt thoughts about self-harm. If you notice any severe symptoms like extreme loss of control, inability to stay in the present, hallucinations, extreme shock, suicidal thoughts, or even an urge to self-harm, they are medical or emotional emergencies. Seek help immediately and don't procrastinate. When left unchecked, it could harm you, as well as those around you.

Chapter Three: Setting Goals for Emotional and Mental Wellness

Have you ever set goals for yourself? If yes, then you are just like everyone else. Have you been able to attain all of the goals that you set for yourself? Probably not, and that is okay. Not a lot of people attain the goals they set for themselves, and it is often because they don't set the right goals. Unless you have certain goals in life, it becomes difficult to stay on track. How can you determine what the right direction is if you are unaware of the end result that you want to achieve? This is where goals come in. A goal could be anything you wish to accomplish.

A goal is an opportunity for change and committing oneself to a course of action. People often come up with different and specific benchmarks for targeting their health goals like, "I know my diet will be healthier if I can include one portion of green vegetables in my daily meals," or "I know my stamina has increased when I can exercise twenty minutes longer than usual." These goals are extremely specific, and it does allow you to gauge any progress you make. However, what about your mental health? How can you set goals for something too large to comprehend or understand? It is not just about setting goals, but the goals you set must be attainable as well. If you don't understand your goals or don't have the motivation to follow through, it is pointless. Setting a good goal determines your rate of success. Here are a couple of common goals people often come up with when it comes to their emotional or mental wellbeing:

- I feel sad, and I want to be happier.
- I need to stop stressing out all the time and learn to be more relaxed.
- I must work on improving my self-confidence and self-esteem to feel better about myself.

Well, there is nothing wrong with these goals. Once you verbalize the goals mentioned above, it helps to focus your energy and sets the stage for sufficient self-reflection and growth. However, these statements are rather broad, and it can be a little tricky to figure out where you must start, what to do, and how you can achieve your goals. Moving forward toward attaining your goals must be done in a manner that's realistic, sustainable, and healthy. In this section, let us look at simple tips you can follow while setting goals for your mental and emotional wellness.

Think Differently

Perhaps the simplest tip is to ask yourself how you can live your life differently in order to attain your goal. This kind of thinking enables you to move toward your ideal state of being, instead of worrying about moving away from your existing state of being. For instance, if you want to stop feeling depressed and want to feel happy, start thinking about how differently you might engage in life in order to be happier.

Here are a couple of questions you can ask yourself:

- "If I want to improve my self-esteem and self-confidence, what kind of assertions must I make? What must my internal self-talk sound like if I want to feel better about myself?"
- "If I want to feel happier, what are the different things I must be doing? Are there any specific experiences I should engage in to feel happy?"
- "If I want to feel more relaxed, what should I do more often? What is the kind of mindset I must assume to feel relaxed in life?"

Concrete Answers

Once you identify the different questions associated with your mental or emotional health in the previous step, it is time to look for some answers. Think about all the specific actions or behaviors you can use to attain your goals. For instance, if your goal is to feel happy, then perhaps you can start spending more time with your loved ones. Maybe you can also make it a point to interact more with others and attend at least two social gatherings weekly.

If you are trying to improve your self-confidence, then maybe you must concentrate on improving yourself. Concentrate on your self-talk and make it more positive and affirming. For instance, if you feel like you haven't accomplished anything or are feeling helpless, you can replace such self-talk with, "I am proud of the way I dealt with difficulties in my past, and I have come a long way." Start replacing all your negative thought patterns with positive and desirable ones, and you will automatically feel better about yourself.

Time Limit

Regardless of the goals you set for yourself, there must be a time limit. Give yourself a specific period within which you can track your goals. Once you have set a time frame, it becomes easier not just to track your progress but to measure your progress too. For instance, if you are working on feeling happier, then you can track the number of social outings you attend in a week. Maybe you can start thinking about all the times in a day when you consciously diverted your thinking from something negative to positive. If you feel like you need to put in more effort, then you can do so. Spend some time and think about all the different things that are preventing you from attaining your goals.

If you routinely check on the progress you make, the chances of attaining your goals will increase. Start concentrating on certain areas of your life that will enable you to accomplish your goals. Think of ways in which you can savor your accomplishments and enhance your sense of self-esteem. You must motivate yourself to keep going. Unless this motivation comes from within, the chances of attaining any goals in life will decrease.

Overall Progress

Once in awhile, take a step back and look at your overall progress. Think of it as a self-analysis report reflection. It enables you to see the larger picture and make any changes as and when required. A simple question you can ask yourself in this situation is, "How consistently have I been working toward my goals, and how do I feel about the changes I have made in life?" Once you start looking at the progress you make, you will probably identify areas that are draining you of your energy. Perhaps you stumble across certain personality traits you didn't know you had, which are preventing you from attaining your goals. Maybe your goals need to be revised based on your current lifestyle or situations in life.

There must be some space for reflection to understand how you feel about the different changes you make to attain your goals. It gives you an invaluable opportunity to revel in your accomplishments and allows you to identify any areas in your life where you haven't been able to attain your goals. At times, there might be some obstacles that stifle your progress. By making a note of all such obstacles, you can come up with solutions to overcome them. Or maybe change your course of action to avoid them altogether. While you are trying to make certain changes, you might even identify significant barriers in your thinking that prevent you from excelling in life. The good news is that you can significantly improve your emotional and mental wellness by using DBT. In fact, during self-reflection, you might realize that the biggest obstacle in life is the way you think and feel. Once you learn to regulate all this, attaining any goal will become easier.

Set SMART Goals

A lot of people set goals without realizing their importance. The goal might be benign in the beginning, but it can turn dark when people shoot for the moon and overpromise on certain goals they set for themselves. The mantra: "Shoot for the moon. Even if you miss, you'll land among the stars," doesn't always work when it comes to setting goals. If you set unattainable goals for yourself, you are merely setting yourself up for disappointment. It is okay to aim high, but the goals you set must be attainable. If not, you merely increase self-doubt and the disappointment you feel in life. In your bid to attain things quickly, you are sabotaging any progress you make. People often expect that they will experience a quick burst of energy whenever they set goals for themselves. Unfortunately, this kind of thinking does not lead to success. If you wish to make solid and lasting progress, it takes plenty of time, effort, and patience. It also requires resilience to bounce back from certain setbacks, which are inevitable. Regardless of the goal you set, these are certain traits you cannot do without.

A common health goal is, "I want to lose weight." It will become incredibly difficult to attain this goal if you aren't sure how much weight you want to lose, by when, and why you want to lose some weight. Unless you understand the reasons for your goal, it becomes tricky to maintain your motivation to attain it. The details involved in setting a goal, especially the minute ones, cannot be overlooked. Now, you might be wondering: why does it seem easy to set goals for physical health, but not mental health?

The acronym SMART stands for goals that are simple, measurable, attainable, realistic, and time-bound. Here is a simple example of how to use the acronym SMART while setting goals for your mental health. A man called Adam has lived with anxiety for as long as he can remember. Due to his anxiety, he always had issues scoring well on tests in school, despite being a good student. He managed to get a job, and once he did this, all the responsibilities he had to deal with increased. He started experiencing general anxiety while interacting with his manager, and then full-blown panic attacks because of work pressure. All this prompted him to come up with a plan to improve his mental health and manage his anxiety effectively. Here is how Adam set goals while using the SMART technique:

Since Adam's goal was to reduce anxiety, he started thinking about the different tools he could use to reduce his anxiety and deal with panic attacks. By coming up with specific techniques, he started to feel more confident while at work. To make his goal measurable, Adam decided that he must track his emotions daily and started rating the anxiety he felt on a scale of one to ten. He tried to do this at least once daily to get a better understanding of how he felt, and the different events that triggered his anxiety. Adam wanted to feel less anxious, and this is a simple yet achievable goal. After a little research and meeting with a therapist, Adam realized his goal was reasonable and perfectly attainable. Therefore, he believed with treatment, effort, and patience, he could realistically attain his goal of reducing his overall anxiety. The final step Adam followed while setting his goal was to set a time limit. He wished to work on improving himself to reduce anxiety before the year ended. It gave him a specific time limit within which he could implement various strategies for handling his anxiety.

As you can see, Adam has managed to set a realistic and well thought out goal for himself. This goal is not only measurable, but it also comes with certain markers he can use to track his progress. Since it is time-bound, it makes the goal more realistic and tangible. So, while setting a goal, you must also follow the same steps that Adam followed to make your goals attainable.

Regardless of the goal you set for yourself, attaining it takes motivation and commitment. Don't push yourself too hard or try to achieve your goals quickly. Instead, start making them more realistic for yourself. You don't have to do things because someone else is doing it. Even if the goals you set for yourself don't make sense to others, don't worry. As long as your goals are in sync with your ambitions and are not unrealistic, you can attain them. Even if you falter, you don't have to worry. Everything is a learning process. Every mistake you make teaches you a lesson. So, stop worrying and instead concentrate on setting SMART goals for yourself.

Chapter Four: Anxiety Disorders – 8 DBT Techniques for Instant Relief

Use DBT for Anxiety

Sigmund Freud classified anxiety into two categories—appropriate and inappropriate stress. Anxiety occurs in any situation that triggers the "flight or fight" response in your brain. When this happens, your limbic system, controlling the sympathetic nervous system, goes on the override and causes symptoms like feelings of nervousness, panic, shortness of breath, or rapid heartbeat. These symptoms can occur regardless of whether the threat is real or not. Since your body cannot distinguish between imaginary and real threats, the defense system it uses stays the same. For instance, even if you're not in a life-threatening situation, but are worried about completing certain tasks at work, your body assumes it is under stress. In this situation, it triggers its flight or fight response. That, in turn, creates anxiety.

Emotions play a vital role in our lives. Basic emotions like fear or stress are associated with anxiety. In a life-threatening situation, these emotions make sense, since the fear motivates us to protect ourselves. At times, these emotions crop up when they are unhelpful or unproductive. They can become difficult to deal with and end up causing anxiety or extreme distress. DBT helps work through your emotions using cognitive skills and applying those skills to your life in general. It helps tackle any difficult or distressing emotions while improving your ability to regulate them. That, in turn, gives you better control of your emotions, including how you experience and express them.

The primary goal of DBT is to change and influence one's emotions. However, before you can do this, you must understand and know where these emotions come from, and the reasons why they crop up. DBT provides non-judgmental and mindful techniques for the observation and description of any emotional experiences you have.

Basic DBT Skills

In this section, let us look at some basic DBT skills. These skills can be used not just to tackle anxiety but to improve your overall emotional and mental wellbeing too. Keep in mind that it takes plenty of time, effort, and consistency to develop these skills. Once you get the hang of it, you will notice a positive change in your life. You will also be more in control of your emotions. Anxiety occurs when you cannot control your emotions and are overwhelmed by them. By learning to regulate them, you will have a better sense of your life and come up with a rational response instead of an involuntary reaction.

Mindfulness

Mindfulness is a simple technique of living your life in the present moment without allowing it to get hijacked by thoughts of your past or the future. You can become more aware of your feelings, thoughts, behaviors, and reactions by being mindful. Mindfulness enables you to take a step back and check-in with yourself, identifying any emotions you feel, and then make conscious decisions based on these emotions. Mindfulness comes in handy in different aspects of your life and is not just a technique for dealing with anxiety. When left unchecked, anxiety can prevent you from leading a happy and fulfilling life. It is an undesirable emotion, and avoiding or regulating it must be your priority.

The easiest way to practice mindfulness is by concentrating on your body. When was the last time you checked in with yourself? Did you ever ask yourself why you are feeling certain emotions? Did you take the time to understand how your emotions work? You can go for a walk and practice mindfulness the whole time—notice how your body feels as you walk. Pause and look at your surroundings. Take in the sights of nature and notice how every movement you make feels. Whenever your mind starts to wander, redirect yourself to the present. You can choose to focus on your external environment or concentrate on your internal experiences. To redirect your mind, you can concentrate on whatever is happening around you, or focus on your emotions, thoughts, and any other physical sensations. The only thing you must do is live the experience in the present.

For instance, if you are lost in your thoughts, and are worried about an upcoming meeting that seems to be the source of your anxiety, then try mindfulness. You might be thinking about all the things that can go wrong or all the tasks you must accomplish for the meeting. All this can be rather overwhelming. Therefore, take a break and mindfully become aware of your thoughts. Keep in mind that whatever you think is just a thought, and it isn't real, at least not yet. You have the power to change the course your life takes. Learn to accept and understand this, and you will feel better. To do this, you must effectively stay in the present without allowing your thoughts to run wild.

Don't Judge

We are often extremely critical of ourselves. This kind of self-criticism is desirable because it offers a chance for improvement and redemption. However, when left unchecked, excessive self-criticism induces plenty of anxiety. Once you start doubting yourself, your abilities, and everything you have accomplished in life, it becomes incredibly difficult not to feel hopeless. Anxiety in such situations can prevent you from seeing any opportunities present right before your eyes.

At times, we also become critical of ourselves because of our thoughts. If you want to deal with anxiety, then you must be non-judgmental. Take a non-judgmental stance in life, and you can deal with anxiety. Perhaps you are used to judging things as being either right or wrong and good or bad. Whenever you start making any negative judgments, you are merely increasing your emotional pain. So, whenever you feel angry or frustrated, notice the judgments you are making. Then, consciously focus on replacing the judgment with facts and emotions you feel.

For instance, if you are feeling anxious because you have several tasks to complete, and you don't think you can complete them on time, instead of allowing your anxiety to trigger negative emotions like frustration, anger, or even crippling self-doubt, it is time to take a pause. Instead of judging yourself critically, change the thought itself. Perhaps you can tell yourself, "I know I have a lot to accomplish, but I can get it done on time," or "If I break the tasks down and make a list, I can complete as much as I possibly can before the deadline." By merely changing your view of a situation, you'll be better equipped to deal with it.

Acceptance

Accepting your reality is a great way to deal with anxiety. There might be certain things you are unhappy with or some painful events that are the source of emotional distress. Unless you accept the pain you feel and acknowledge whatever has happened, you cannot move on. If you allow your emotions to overwhelm you, then they will quickly overpower you, and you will feel out of control. Instead, start by accepting them. Let us look at the previous example and use acceptance to make this situation bearable. The reality you must accept in the situation is that you have plenty of work to complete. You cannot deal with the task at hand unless you accept the situation. Once you accept it, you can think clearly and calmly. Once you are calm, you can come up with solutions to deal with the situation instead of worrying about it. By simply accepting your reality, you can reduce the stress you feel and avoid unnecessary anxiety.

8 DBT Techniques for Instant Relief

Technique #1: Understand the Source

If you want to deal with your anxiety, unless you identify the source or reason, it becomes difficult. Take a while and think about any specific situation that triggered your anxiety. Perhaps you could have avoided it by saying no. Or maybe by asking yourself what you're supposed to do. Once you identify the source, you can take immediate action to rectify it.

Technique #2: Concentrate on Breathing

The simplest technique you can use for immediate relief from anxiety is to concentrate on your breathing. Take a break, and focus on nothing else other than your breathing. Inhale and exhale deeply. While doing this, notice how you feel and don't concentrate on any of your thoughts. You can concentrate on your breathing for up to ten minutes, and you will immediately feel better.

Technique #3: Self-Soothing

We all experience life using our five senses—vision, touch, hearing, smell, taste. If you feel stuck at work, take a break and head outdoors to enjoy nature. Stop whatever you are doing and listen to some calming music. You can drink herbal or flavored tea or eat a small treat to stimulate your sense of smell and taste. Applying scented lotion, inhaling the beautiful smells of nature, or lighting a scented candle will make you feel relaxed. If possible, wash your face with cold water, or when at home, you can take a warm bath. If you have a pet, play with your pet. Learn to self soothe, and your anxiety will slowly crumble away.

Technique #4: Practicing Mindfulness

Mindfulness is a very simple exercise that allows you to process your thoughts and emotions rationally. Whenever you feel overwhelmed, take a break and shift your attention to whatever you feel. Allow yourself to feel your emotions without any judgment. Mindfulness allows you to stay in the present without worrying about the past or future. When you do this, you'll get a better sense of yourself and the situations you have to deal with. It also helps regulate your anxiety.

Technique #5: Exercise

Exercise is not just good for your physical health but your mental health as well. Whenever you exercise, your body produces stress-busting and mood-boosting hormones like endorphins. It helps counteract the harmful effects of cortisol or other stress-producing hormones in your body. A quick jog, jumping in the same spot for a couple of minutes, or doing anything else that gets your body moving will help alleviate stress.

Technique #6: Stimulating Activities

Engage in different activities that stimulate your mind and require thinking. Solving puzzles, word searches, or reading will help shift your focus from anxiety to something more positive and constructive. By diverting your attention toward other activities, you can prevent your mind from dwelling on your anxiety. Plan for a couple of fun activities you can engage in regularly. By spending time with your loved ones or by working on your hobbies, you will feel better.

Technique #7: Help Others

At times, it helps to help others. Instead of thinking about yourself, take a moment and try to think of how you can help others. Contributing to society is a great way to tackle anxiety. It gives you a chance to do something nice for others and to stay in the present.

Technique #8: Your Accomplishments

Anxiety can be crippling. It can induce feelings of self-doubt. So, give yourself a break, and think about all of your accomplishments. It doesn't have to be anything significant, and even small victories matter. Instead of dwelling on all your mistakes or all the things that have gone wrong, concentrate on the good you have in life. Concentrate on your accomplishments and be grateful for them.

By following the simple tips given in this section, you can get instant relief from anxiety and stress. Apart from this, start working on the three basic DBT skills discussed in this section to alleviate anxiety and regulate your emotions.

Chapter Five: Depression and Emotional Regulation – 7 DBT Tips to Feel Better NOW

Common symptoms associated with depression include prolonged periods of sadness, an inability to focus or concentrate, reduction in your memory power, withdrawal, and a lack of interest in all activities you formerly enjoyed. Depression can also manifest itself as physical symptoms like lethargy, inability to sleep, headaches, and body aches. While experiencing these overwhelming feelings associated with emotional pain, overcoming the symptoms of depression can be tricky. DBT enables you to accept yourself along with your current situation so that you can take corrective action. By using DBT, you can effectively overcome the common symptoms of depression.

DBT helps you cope with painful emotions and come up with effective coping skills to overcome any challenges you face. Finding the ideal balance between change and acceptance is essential for emotional regulation. The positive reinforcement provided by DBT can allow you to overcome any crippling feelings of hopelessness or acute sadness. DBT helps with emotional regulation.

Controlling your emotions, keeping them balanced, and avoiding them from going to extremes is emotional regulation. Emotional dysregulation is your inability to control your emotional responses. Usually, an internal or external event triggers a subjective experience like an emotion or feeling. This emotion or feeling results in a cognitive response or a thought that is followed by a physiological response based on emotions. When you are thinking about something sad, it promotes negative thinking and results in a physiological response like an increase in your heart rate or production of stress-inducing hormones. This process often culminates in undesirable behavior, which could be in the form of avoidance or extreme expressions of emotions.

Emotional dysregulation is often characterized by emotional reactions that are exaggerated. Even a relatively small negative event warrants an exaggerated and over the top emotional response in an emotionally dysregulated individual. If you feel like screaming, crying, or have a mental breakdown whenever a small inconvenience presents itself, it is a sign of emotional dysregulation. Blaming others or displaying passive-aggressive behavior can effectively harm any of your existing relationships and increase the conflict in the situation.

A great thing about DBT is that it concentrates on practical skills and techniques with real-world applications. In this section, let us look at simple tips you can use for managing depression and promoting emotional regulation to feel better immediately.

Tip #1: Identifying Emotions

The easiest way to regulate your emotions is by identifying and labeling them. DBT encourages its users to come up with innovative and descriptive ways to label their emotions. Instead of using generic or regular terms, the primary idea is that unless you know what the emotion is, you cannot manage it. In DBT, you will also learn the difference between primary and secondary emotions and how to address each of them effectively.

Primary emotions are the initial reaction to a trigger or an event and tend to be natural reactions, while a secondary emotion is a reaction to the primary one. For instance, you might be angry when you have an argument with someone or upset when you don't get something you want. Secondary emotions are more dangerous than primary ones. However, you can regulate your secondary emotions, and they are well within your control. It is entirely up to you how you react when you argue with someone. Your secondary emotions might trigger self-destructive or maladaptive behaviors. Unless you learn and accept your primary emotions, you cannot regulate your secondary emotions.

While understanding and labeling your emotions, keep in mind that there is no such thing as a right or wrong emotion. Emotions are normal, and the only thing that matters is how you react to them. Your emotions are unique, and you don't have to adhere to any notions of what others think about desirable and undesirable emotions.

Tip #2: Let Go of Undesirable Emotions

If an emotion doesn't lead to your growth or harms you in some way, then it is undesirable. Learning to let go is a skill that comes in handy in different aspects of your life. We often get stuck in a loop of negativity while processing negative emotions. In an attempt to understand these emotions, we hold onto them tighter than we are supposed to and start obsessing over minute details associated with the emotional experience.

Even if it sounds paradoxical, acceptance is the first stage of letting go. Unless you accept the emotion you are feeling and acknowledge the fact that you don't like it, you cannot let it go. You must stop running away from your emotions, and instead face them head-on, once you accept your suffering. Acceptance also brings with it a sense of clarity. Start by observing a feeling, acknowledge it, and let it go. Understand that your feelings are just a part of you, and they do not define you. At times, the best thing you can do is avoid reacting to an emotion, and just allow it to stay with you for a while. Usually, reactions to a situation prolong suffering. Start learning to love and accept all your emotions unconditionally. Acceptance is not the same as approval. Even if you don't approve of a feeling, accept it, and only then can you let it go.

Tip #3: STOPP Technique

STOPP is an acronym that stands for stop, take a breath, observe, pull back, and practice. This technique works brilliantly well while dealing with any intense reactions. It prevents you from reacting in the heat of the moment and instead promotes mindful and rational reactions. It is also a form of mindfulness. Whenever you feel an intense emotion, stop for a moment. After you take a pause, take a deep breath and exhale slowly. These are the first two steps.

The third step is to observe all the thoughts going through your mind and understand where your attention lies. Think about what you are reacting to and the different sensations you experience in your body. The fourth step is to put things in perspective and look at the bigger picture. Perhaps the view you have is not the only perspective, and there could be another paradigm you haven't considered. Think about how your loved one would react in a similar situation and discern whether you think it is an opinion about how they would react, or a fact. Alternatively, you can also think about how important the specific issue will be in a couple of months. It enables you to understand whether your perspective is desirable or not.

In the fourth step, start practicing whatever works for you. Think about the best course of action you can take. Think about whether your action fits well with your values or not. The only way you can master this skill is through constant practice. Whenever you give yourself a break between an intense emotional reaction and the subsequent action, you can effectively regulate your emotions.

Tip #4: Opposite Action

Whenever you experience an intense emotion and wish to stop it, use the technique of the opposite action. Emotions are always associated with a specific behavior—like an argument that triggers anger, or you might experience an urge to withdraw when you feel sad. However, most of us wrongly assume that emotions trigger behavior instead of it being the other way around. You can trigger an emotion by engaging in a specific behavior associated with that emotion. Instead of behaving how you normally do when you feel a certain way, opt for the opposite course of action. If you yell whenever you are angry, try to talk quietly. When you feel sad, talk to your loved ones instead of isolating yourself.

Tip #5: Check Facts

Placing excessive importance on your emotions or blowing things out of proportion is quite easy. By checking facts, you can effectively identify a scenario when it occurs and then work on reducing the intensity of your emotional response. To check the facts, here are certain questions you can ask yourself.

- Did anything specific trigger my emotional response?
- What are the different interpretations I have about an event?
- Does the intensity of my emotional response match the intensity of the situation?

Tip #6: DEAR MAN Technique

DEAR MAN is an objective effectiveness skill taught in DBT. Here are the different steps you must follow while using this acronym.

Describe

While dealing with overwhelming emotions, start by describing clearly and precisely what you want. Be polite while you describe what you desire from the other person. For instance, if you get annoyed that your partner doesn't clean up after cooking, maybe you can say, "Can you please do the dishes after cooking?" Instead of, "You never clean up."

Express Yourself

Others cannot read your mind, and you must not expect it also. Unless you express yourself clearly, you cannot expect others to comprehend what you want. While expressing your feelings about the situation, stick to "I feel" sentences instead of "You don't." Let us go ahead with the example discussed in the previous step. Instead of saying, "You don't ever listen to me and you never clean up," you can say something like, "I feel frustrated because it feels like you aren't listening to me."

Assert Yourself

Stop beating around the bush and get to the point. For instance, if you think you don't have the energy to cook, instead of saying, "I don't think I will be able to cook tonight," you can say, "I don't have the energy to cook tonight because I have lots of work to catch up on." Be assertive and express yourself clearly and without any ambiguity.

Reinforce Good Behavior

If others respond well to you, you must not forget to reward them. It also helps to reinforce why the outcome you desire is positive. It could be something as simple as a thank you or a polite smile. For instance, if your partner does do the dishes and cleans up the kitchen without any reminders, don't forget to acknowledge the effort they have made. Tell them you appreciate it and reinforce why it is a good idea to always clean up.

Mindfulness Matters

While in the midst of an argument or a heated debate, it becomes easy to get sidetracked and forget about the objective of the interaction. Be mindful of all that you say, and always stick to the issue at hand. Don't get distracted, and don't combine all your issues into a single argument.

Appear Confident

Appearances matter and you must seem confident. Your body language and communication must display confidence. Unless you appear confident, the likelihood of others listening to you will reduce.

Negotiation

It is highly unlikely that you will always get whatever you want out of all of your interactions. Therefore, learn to negotiate and be open to all negotiations. For instance, if you want your partner to do the dishes, maybe you can offer to do something in return. It will work as a motivation to convince him to do the dishes.

Tip #7: FAST Technique

If you ever find yourself in a situation where you are struggling to stick to your ideas, then use the FAST technique. This technique will help improve your self-respect and effectively assist in regulating your emotions. Here are the different steps you must follow while using the FAST acronym.

Fair

Learn to be fair, not just to others, but yourself as well. You deserve the compassion you extend toward others.

Apologize

Never apologize for making a request or sharing your opinions. You are entitled to disagree and don't allow anyone to make you feel guilty for it. Apologize when you make a mistake, but not unnecessarily.

Stick to Your Beliefs

Just because you want something, don't compromise on your values or beliefs. Your values and beliefs define you and your thinking. Always stand up for what you believe in, especially when you know you are right.

Truth Matters

Avoid any forms of manipulation like an exaggeration, lying, or putting on the act of a victim/martyr. Always be truthful and avoid dishonesty at all costs.

By using this technique, it will serve as a quick reminder to ensure that you stay on the right track without indulging in any unnecessary negative thinking.

Chapter Six: Workplace Stress: 9 Ways to Use DBT at Work

Workplace stress has become incredibly common these days. Our unsustainable ideas of success have resulted in a culture of excessive burnout, and it is harming our ability to be creative, efficient, and productive. Multitasking has become the norm a lot of people follow. However, all these things are merely increasing the stress we feel. Different external and internal factors contribute to stress. You might feel like getting through a usual workday is becoming incredibly difficult. If this is true for you, then you are not alone. If you wish to reduce the stress you experience, then it is time to change the way you think about work. By changing your perspective, you can reduce the amount of anxiety and tension you feel while at work.

If you experience any difficulty regulating your emotions, especially in a place where you must display professionalism, it can be detrimental to your growth. Tensions usually run high at workplaces. Stress is quite common, and keeping your emotions in check can become difficult in such instances. Some of the common conflicts you might face at work include dealing with hierarchy, confusing and cumbersome protocols, feeling as if your opinions don't matter, complicated relationships with co-workers or superiors, and so on. The list of conflicts you can probably run into in the workplace is endless. If you don't deal with them effectively and efficiently, it will take a toll on your mental and overall health. Dealing with workplace stress is a skill that anyone can use.

Whenever something occurs, our interpretation of the facts or the course of events is guided by our emotions. What we feel is not often the result of the facts we know, but it is about what we think. All the interpretations we make about different events in our lives determine how we feel in general. DBT allows you to take a step back from the situation and carefully analyze the task at hand to make rational and efficient decisions. To do this, you must learn to regulate your emotions, and the workplace is not an exception. One of the main areas that DBT concentrates on is mindfulness. Unless you are mindful of yourself and your surroundings, you cannot regulate your emotions. DBT also teaches you acceptance and how to regulate your emotions.

If you are tired of allowing the stress at work to get the better of you, or you want to improve your ability to manage stress, then here are some DBT techniques you can use:

Technique #1: A Sacred Space

Take a break whenever you feel like your focus is wavering, and your mind is overrun with thoughts of worry and stress. Step away from your desk for a couple of minutes and allow yourself to center your focus. Taking a quick break and removing yourself from a stress-inducing situation gives you a chance to reset your thinking patterns. Make it a point to take a break after working for an hour. It could be something as simple as getting up from your desk and refilling your water bottle, making a cup of coffee, or even going outdoors for ten minutes. To calm yourself, concentrate on your breathing. All it takes is five minutes to reset your mind and let go of any stress you feel. Concentrate only on your breath and visualize positive imagery or listen to cheerful music.

Technique #2: Start Prioritizing

As the demands keep increasing and the daily tasks start adding up, your stress levels will increase. The easiest way to tackle the stress is by prioritizing all the tasks you must complete. While prioritizing, arrange tasks in the order of their importance. Concentrate on the tasks that are extremely important and leave the rest for later. Once you remove all the unimportant tasks, it helps to reduce your stress as well. Usually, while working, we give in to the illusion that we can complete all the things that we have to do in the given timeframe. It is okay to believe in your abilities to get things done, but it is never okay to overestimate your productivity. If you aren't careful, you will end up in a situation where you have bitten off more than you can chew. In your attempt to try and complete as much of your work as possible, while adhering to deadlines, your stress levels will increase.

The first step is to carefully reassess all the tasks you must complete and your productivity. Make a list of about ten to fifteen things you wish to accomplish within a day and rate them according to their importance. In the list you have created, not all tasks are equal, and some will be more crucial than others. You will end up with a list that includes about four to five important things and four to five unnecessary things. Once the list is in place, it is time to eliminate all the unnecessary tasks. Accept the fact that you cannot complete everything and make your peace with it. Don't try to be a perfectionist if you don't want to deal with the consequences of burnout. Focus only on the important tasks, and it becomes easier to tackle them.

Technique #3: Monitor Your Mood

Here is a simple technique you can use to recognize and change any of your negative thought patterns. Take a break for two minutes, grab a pen and paper, and divide the paper into three columns. In the first column, make a note of any stressful event, write down your feelings in the second column, and rate them on a scale of 1-100 in the third column. For instance, if the stressful event is that you are worried about a specific meeting, make a note of it. In the second column, write down any feeling you experience in a single word, such as anxious, unprepared, overwhelmed, or scared. In the third column, rate this event with 100 being extremely overwhelmed. While rating the event, think about all the thoughts going through your mind. Once you start making a note of everything you feel, it becomes easier to monitor and reassess your mood. Once you have completed this exercise, fold the paper, and forget about it for one day. The following day, you can reevaluate whatever you have written down, and you will realize your worries were usually unjustified.

When you practice this technique over a period, you will realize certain negative thought patterns. Once you identify them, it becomes easier to deal with them.

Technique #4: Probability Works

There will be several instances at work when you feel like you are not doing enough or that you will not be able to complete certain tasks assigned to you. You might end up feeling suffocated because you start worrying about all the things that can go wrong. Thoughts like, "This project will not be completed on time. My manager will hate me. I will probably get fired. I will not have a job" can effectively halt your productivity. This kind of thinking is known as catastrophizing. Most of us are guilty of indulging in a little catastrophizing at one point or another. When you start spending all your mental energy concentrating on a possible failure or hurt, you increase the stress you experience.

When you view the world as black-and-white and assume a success or failure mindset, your anxiety levels increase. Keep in mind that things don't always work out as we want them to, but it doesn't mean it is the end of the world if they don't. Avoid all these destructive thoughts by using probability. Whenever you notice your thoughts are spiraling out of control and you are feeling stressed, take a moment to think about the situation at hand. Usually, things are seldom as bad as we imagine them to be. For instance, if you're worried that a project will fail, take a moment, and think about the likely outcomes. If the probability is one out of ten—or even four out of ten—then wasting any time thinking about the worst-case scenario is not productive. If the chances of the worst-case scenario are quite low, then why worry about it? When you reframe your perspective about a situation, you get a chance to make a conscious choice. You can worry about the situation for as long as you want, but no good will come out of it. Instead, get back to the task at hand and try to do your best. Just because you did not succeed at something, doesn't mean you failed. It merely means there is scope for improvement.

Technique #5: Avoid Multitasking

Don't be under the illusion that multitasking means you can get more things done, and that it will improve your productivity. In fact, it is counterproductive. When you multitask, you merely increase the stress you feel. Whenever you feel a little overwhelmed, concentrate only on one thing. If your mind starts thinking about other tasks, merely re-center yourself and get back to the task at hand. When you start completing one task after the other, you can accomplish more in a day, and it reduces stress.

Technique #6: Tune out Distractions

While working, get rid of all distractions. Place your phone on silent; avoid talking to your co-workers while working; and log off your social media. If you want, you can also set a timer for 45 minutes or an hour. When you tune out all distractions, your productivity increases, and your mind will not think about unnecessary things. If you keep practicing this technique, you will notice that your overall productivity increases, while your stress decreases.

Technique #7: Concentrate on Yourself

Take a break and concentrate only on yourself for three minutes. Close your eyes, connect with your body, and understand what you are feeling. Acknowledge your thoughts, and don't dwell on them. Allow them to pass through your mind freely without any judgment. Check-in with your senses and let go of any tension present within your body.

Technique #8: Gratitude

Expressing your gratitude for all the good things you have in life will have a positive effect on your health, creativity, and working relationships. Instead of worrying about all the things that can go wrong, concentrate on the good you have in life. Take a couple of minutes to think about all the people who have supported and helped you throughout your life. Be grateful for the support, and thank the universe for all you've been given. When you concentrate on the good things in your life, your mind will not dwell on any negativity. Instead of worrying about all the projects that are due, think about all that you have accomplished, and it will instantly lift your spirits. When in doubt, remind yourself of all your accomplishments, and you will feel better.

Technique #9: Exercise Break

Schedule an exercise break for yourself. You can practice simple yoga poses at your work desk or go for a brisk walk. Whenever you head to the bathroom or go to grab some coffee, use this opportunity to stretch your arms and relax your muscles. Bend your head and neck, stretch your arms over your head, and stretch your legs. It allows the blood to flow freely through your body and will make you feel energized.

Chapter Seven: Borderline Personality Disorder: Tame Impulsions and Mood Swings with DBT

The way you think and feel about yourself and others is negatively influenced when you have borderline personality disorder (BPD). It often causes troubles and prevents you from functioning effectively in daily life. It can also cause issues with self-image, results in unstable relationships, and makes it difficult to manage your emotions or behaviors. Borderline personality disorder brings with it an excessive fear of abandonment or instability, and you might experience difficulty being alone. Impulsiveness, inappropriate expressions of anger, and constant mood swings might push others away, even if you want to build lasting and healthy relationships.

Usually, borderline personality disorder presents itself in the early stages of adulthood and might get better with age. You don't have to feel discouraged if you have BPD, and DBT is an effective way to handle this mental health disorder. Here are some of the signs and symptoms of BPD:

• Severe fear of abandonment, which might push you to take extreme measures to avoid any imagined or real rejection or separation from others.

- Bouts of stress-induced paranoia and disassociation with reality that can last for a couple of minutes, or even hours.
- Frequent changes in self-image and self-identity that trigger changes in your values and goals.
- A steady pattern of unstable and intense relationships wherein you might idealize someone one instant and then suddenly believe the said person is cruel or doesn't deserve your love.
- Engaging in risky or impulsive behavior like reckless driving, binge eating, drug abuse, alcohol abuse, or self-sabotaging behaviors.
- Intense and frequent mood swings that can last for a couple of hours to days on end. You can experience intense feelings of irritability, anxiety, or even happiness.
- The relentless feeling of helplessness and emptiness that doesn't go away.
- Indulging in suicidal thoughts or attempting self-injury, usually in response to the fear of abandonment or rejection.
- Intense and inappropriate bouts of anger, losing your temper frequently and quickly, or even engaging in physical fights.

Note: If you notice that you are engaging in any fantasies of self-harm, or are thinking suicidal thoughts, then seek medical help immediately.

DBT is commonly used for treating borderline personality disorder. It is the first type of psychotherapy that has been proven to be effective while treating BPD in regulated clinical trials. It is also believed to be the gold standard of treatments for BPD. DBT has been proven to be effective in reducing substance abuse, alleviating thoughts of suicide, destructive behaviors, and reducing the need for psychiatric hospitalization. Emotion dysregulation is the core problem in borderline personality disorder. It is often caused due to biological risk factors and genetic risk factors, along with an unstable emotional environment in childhood or unhealthy means of dealing with chronic stress.

Learning to regulate your impulses and control mood swings is essential for dealing with a borderline personality disorder. Here are certain tips that will come in handy.

Information about BPD

Start gathering as much information about BPD as you possibly can. Once you understand what this disorder is and how it can affect you, it becomes easier to deal with it. Unless you understand what you are dealing with, you cannot manage this condition. Also, start talking about BPD with your loved ones. Having a strong support system in place increases your ability to cope with and manage BPD. Try to understand that BPD, like everything else, exists on a spectrum. While doing this, don't assume the worst. It always differs from one person to another. The usual symptoms associated with it aren't lasting, and this condition can be efficiently managed. Don't allow the diagnosis to consume you. Even at your worst, understand that it is the BPD that's guiding your actions. Once you acknowledge all this, it becomes easier to deal with BPD.

Carefully Consider Your Decisions

Carefully consider all the decisions you made in the past, especially the ones associated with any relationships in your life. At times, you would have reacted harshly or made some rash decisions, which might have cost you a couple of good relationships. Once you start dissecting your relationship history, you can identify any negative thinking patterns that prompted you to react irrationally. By identifying such patterns, you will be better equipped to deal with similar situations in the future. Also, it is never too late to make amends. If you think you were at fault, immediately offer an apology. It will lessen the guilt you feel, and you will feel better about yourself.

Don't Hide It

You don't have to hide your BPD or try to mask it. Don't let anyone else tell you otherwise. You don't have to fit into the societal mold and follow any societal norms that don't suit you. People are entitled to their opinions, and different people will think differently. Don't try to pretend that everything is fine when you don't feel fine. Embrace the diagnosis of BPD and start learning from it. Once you embrace it, it leads to individual growth. All things that did not make sense will finally fall into place. You can finally understand your actions and realize why your life is the way it is. Start sharing information about BPD with your loved ones and help them understand what you are going through. Once they understand what you are going through, they will be better equipped to deal with your emotional responses. Keep in mind that your BPD does not define you, and it is not your fault. Don't blame yourself, and don't indulge in any forms of negative thinking.

Use a Journal

A simple way to disengage from any emotions you feel and lighten your emotional load is by making a note of your feelings. Maintain an "emotion journal" or diary where you can pour your heart out. At times, you cannot explain things to yourself, or even those around you. In such situations, grab your diary and start writing down whatever you feel. Don't judge yourself for the things you write. Don't try to analyze them. Instead, think of it as an outlet for your emotions. When you look at whatever you have written, you will be in a better position to understand your feelings. Once you understand your feelings, you can react appropriately. It also gives time for self-reflection and self-analysis. Apart from this, writing down your thoughts and emotions is a great way to engage your mind. Instead of worrying about certain things, write about it, and you can come up with appropriate responses.

Understand Your State of Mind

Mindfulness is essential for DBT. Understanding your state of mind will enhance your ability to regulate your emotions and keep them in check. The three states of mind in DBT are: reasonable mind, emotional mind, and wise mind. While in a reasonable mind, you tend to approach things logically, plan your behavior, pay attention to any information available, and react appropriately. It could be something as simple as measuring the ingredients before cooking. The reasonable mind enables you to stay logical and rational.

The emotional mind, on the other hand, is primarily controlled by your emotions. Logical thinking goes out the window, and facts get easily distorted. This state of mind often results in impulsive behaviors and negative thought patterns. For instance, taking an unplanned trip or arguing with someone merely because you both don't agree on a specific topic. By understanding your emotional mind, you can quickly come up with ways in which you are harming yourself unintentionally. When your reasonable mind and emotional mind come together, it is known as the wise mind. It is about a feeling of intuition or a sense that something is or is not right. Perhaps you experience intuition that something doesn't feel right or experience something that cannot be explained using your logical mind.

Whenever you feel an urge to react, carefully consider the state of mind you are in, and you can effectively improve your responses to situations.

Analyze the Information

There are two simple steps you can follow while dealing with BPD, and they are known as "checking the facts." Whenever you experience anxiety or extreme distress, check the facts before you respond or react to a situation. The first step is to identify the emotion you are feeling. The next step is to see whether your emotion is justified by checking all the facts or information available to you about the emotion you are feeling. If you feel like your responses are uncalled for, or you are reacting impulsively, then take a step back and calm yourself before you react. This is an important skill that will come in handy in different aspects of your life.

Positive Self-Talk

If you want to learn to manage your reactions to different circumstances and to respond in a healthy manner, then it is time to challenge any of your negative habits or reactions using positive self-talk. It will certainly take some time and effort before you get the hang of it. By concentrating on positive self-talk, you can alleviate anxiety, improve your attention, and stay more focused. Gently remind yourself you are worthy of all things desirable and good in life. Instead of dwelling on all the negative thoughts you have about yourself or those around you, concentrate on the positive aspects. Everything in life is temporary, and nothing stays constant. Change is the only constant in life. Therefore, even the toughest of times you experience will pass. So, what is the point of wasting your precious time and energy dwelling on all unpleasant experiences when they are just temporary? Remind yourself that the moment you are experiencing right now does not define your past or the future. You cannot control the future, and it is unpredictable. Worrying about it will merely increase the stress you feel. Instead of dwelling on the past, concentrate on what you can do to improve the situation, the next time you face it. It gives you better control over your behaviors and actions, instead of feeling like you are victimized by life.

Start reframing all your negative thoughts using positive self-talk. For instance, if a presentation didn't go well, you might think to yourself, "I am a failure, and I can never get anything right." Instead of ruminating about such negative things, you can think, "The presentation did not go as well as I hoped it would. I can talk to my colleagues and identify the areas where I need to improve myself." By becoming aware of your negative self-talk and replacing it with positive sentences, you can regulate your emotions.

Check-In with Yourself

Taking care of your health is your responsibility. No one else can do it for you, and unless you take care of yourself, you cannot deal with BPD. Anger and despair are usually the natural reactions for any circumstance or situation when you have BPD. For instance, if your friend did something to upset you, your initial instinct might be to throw a tantrum or even threaten the other person. Instead of doing this, take a moment and check-in with yourself. Once you do this, you can effectively communicate what you feel to the other person in a non-threatening manner. By following the simple mindfulness practice, you can prevent relationships from turning sour. Also, you will become better equipped with your emotions and thinking patterns.

For instance, if your partner was late for your date, your immediate response might be to get angry. Your usual response might be to yell at him and ask why he is so inconsiderate or disrespectful toward you. Before you do this, take a moment and check-in with yourself. Identify your primary and secondary emotions. For instance, you are probably angry or upset because you are worried that your partner doesn't care or because of any abandonment issues you have. Once you have identified the emotions, you can ask him why he was late. It certainly is a better way of dealing with the situation.

Chapter Eight: Improving Your DBT Distress Tolerance Skills

There will be instances in your life where you feel extremely distressed. Perhaps it is impossible or extremely difficult to change the scenario you are dealing with. In such instances, what can you do? Well, this is where DBT comes in. DBT teaches certain distress tolerance skills that enable you to cope with and survive a crisis. It helps you tolerate short-term as well as long-term pain, either emotional or physical. In this section, let us look at certain simple tips you can follow to improve your DBT distress tolerance skills.

ACCEPT Technique

ACCEPT is a DBT technique that includes a group of skills; you can use it for tolerating any negative emotion until you can resolve the situation. ACCEPT stands for: activities, contributing, comparison, emotions, pushing away, and thoughts. Whenever you feel distressed, start by diverting your mind by indulging in different activities. Try to keep yourself busy so that you don't dwell on any negative emotions. You can do the dishes, go for a walk, read a book, get some work done, or even engage in any of your hobbies. Instead of wasting your time thinking about an unproductive task, concentrate on doing something productive.

Try to do something for others instead of getting overwhelmed by your own emotions. Whenever you contribute to someone else's wellbeing, it enables you to stop thinking about your problem at hand. Perhaps you can bake cookies for a loved one, help a neighbor with certain chores, or even cook dinner. It is easy to lose perspective while going through any distressing times. Were there any instances in your past when you thought you wouldn't be able to get through something, but you did? Remember such instances and tell yourself you can also get through the situation at hand. It certainly puts things in perspective and gives you the internal motivation to hold on tight.

You can always regulate your emotions regardless of the situation. If you are feeling distressed, you have the power to experience the opposite emotion. Whenever you feel anxious, start meditating for about ten to fifteen minutes. If you feel depressed, talk to one of your loved ones. If you feel sad, then concentrate on a hobby you enjoy.

Whenever you feel like you cannot deal with something, temporarily tune it out of your mind. Temporarily push that thought away by distracting yourself with other activities, mindfulness, or other thoughts. You can give yourself a break and then get back to the issue once again when you feel ready and better. Replace all the negative thoughts with activities that excite your mind and distract it. Maybe you can try saying the alphabet backward, count backward, or solve any puzzles. It enables you to avoid any self-destructive thought patterns or behaviors while regulating your emotions.

Radical Acceptance

There will certainly be times in your life when you get stuck in undesirable situations incapable of any change. You might vehemently hate it or disapprove of it, which only increases the distress you feel. If you merely accept it for what it is, you can reduce the distress you feel and be at peace with yourself. Once you stop dwelling on the issue or circumstance, you can easily move on.

Radical acceptance is a simple skill that believes we all tend to have choices, and at times, the only choice you might have is to decide whether you wish to accept the reality of a situation or not. You have the choice to fume over the issue and stay miserable because of it. Or you can decide to accept it and move on. For instance, let us assume that you have a cavity but are terrified of going to the dentist. You might try to ignore it, avoid it, and even deny its existence. You liked the previous dentist and had a good relationship with him, but he retired. The new dentist doesn't seem understanding, and you don't like them. In an attempt to avoid going to the dentist, you start eliminating your favorite foods that irritate the cavity, like all sugary treats. It works out well for you since you are eating less unhealthy foods now. However, once in a while, the cavity flares up, causing unbearable pain.

If you practice radical acceptance, it gives you the strength required to accept that you are scared of going to the dentist and that it will probably be an unpleasant experience. However, it also gives you the courage to accept the fact that you must get the cavity filled. With radical acceptance, you accept the worst and get through the experience to move on. Learning distress tolerance skills is certainly not easy, but it is desirable.

TIPP Technique

Maybe you have reached your emotional breaking point. It could be a time where the saying, "The last straw that broke the camel's back," starts making sense to you. In such instances, TIPP is the DBT skill that comes in handy. TIPP is an acronym that stands for: the temperature of your body, intensive exercise, pace your breathing, and progressive muscle relaxation.

Emotions tend to manifest themselves physically. For instance, whenever you feel upset, your body might feel a little warm. To counteract this, wash your face with some cold water, turn on the AC and sit close to it, or you can also hold an ice cube. By changing your body's temperature, it helps you to cool down—physically and mentally. If you are experiencing intense emotions, try to perform or engage in any intensive exercise that matches the emotion you feel. You don't necessarily have to be a marathon runner. You can start running or jogging a couple of times, or maybe swim until you tire yourself out. Intensive exercise increases the flow of oxygen in your body while counteracting the stress levels.

Emotional pain can be reduced by concentrating on your breathing. There are different types of breathing exercises you can follow to do this. The simplest one is known as "box breathing." To do this exercise, find a quiet spot for yourself and get started. Inhale deeply and slowly through your nose and hold it to the count of four. Now, slowly exhale through the nose to the count of four. Hold onto your breath to the count of four and start again. By focusing on a steady pattern of breathing, it reduces the stress you feel.

Progressive muscle relaxation is quite intriguing. Start by deliberately tightening a muscle for around five seconds, then relax it and allow it to rest. Move onto another muscle and do the same thing. Once you do this, your muscles will feel more relaxed than they did before. When your muscles are relaxed, their oxygen requirement reduces; it slows down your breathing and heart rate. That, in turn, will calm you down. Focus on any group of muscles that feel tense to you like the ones in your arms or shoulders. Consciously relax the muscles, and within no time, you will start to feel better.

List of Pros and Cons

Whenever you feel distressed by a situation, making sensible decisions becomes exceedingly difficult. DBT suggests you can make a list of pros and cons to weigh the consequences of any decision you make. Engaging in behaviors of self-harm or anything self-destructive while dealing with an emotional crisis is quite common. Before you act on an urge, make a list of pros and cons about whether you should go ahead or not. Once you start writing things down, you'll get more clarity and can rationally decide instead of allowing your behavior to be guided by undesirable urges.

IMPROVE Skill

There will be times when you cannot control an unpleasant event, regardless of whether it is big or small. In such instances, you require distress tolerance to make it through without indulging in any unhealthy behaviors. No emotion is permanent, and the same applies to intense emotions too. IMPROVE is a DBT technique that stands for: imagining, meaning, praying, relaxation, one thing at a time, vacation, and encouragement.

Start by imagining how you would feel if you successfully dealt with an unpleasant situation. Try to hold onto the feeling of accomplishment, and you can get through the unpleasantness. The next step is to understand the meaning of a painful situation. What is the lesson you can learn from the unpleasantness you have to deal with? Perhaps it is teaching you to be more empathetic, build new relationships, or it is about healing yourself.

Regardless of whether you are a religious person or not, the power of prayer can never be underestimated. You don't necessarily have to pray to God or a deity, but you can believe in a higher power or the universe. Simply surrender your problem and ask the universe to grant you the power to tolerate the situation. The next aspect of this technique is relaxation. Your body is in a fight or flight mode whenever you experience stress. Engaging in any form of relaxing activity helps relieve the distress you experience. The different activities you can include are walking, breathing exercises, taking a warm bath, or even yoga.

Taking "one thing at a time" is the next step. Mindfulness is about staying in the moment while you let go of any thoughts about the past or future. If you keep adding to old issues, the distress you feel only worsens. Instead of doing all this, concentrate only on one task at the moment. Find one thing that you can shift all your attention to and get it done immediately. You can then effectively take a break from stresses and stressors while on vacation. If you can afford to, take a break and go on holiday. If this isn't possible, then visualize how wonderful you would feel if you were on vacation. Hold onto this positive visualization, and you can enhance any positive feelings you experience.

For the final step, you don't need to wait for encouragement from external sources, it can come from within. Using positive affirmations, or any meaningful phrases can give you the motivation to keep going and make it through difficult times in life. Instead of dwelling on any negative thoughts, replace them with a positive affirmation. Don't think, "I cannot do this, and I am a failure," instead, you can think, "I have got this, and I can get through it."

Learn to Self-Soothe

In a crisis situation, you can effectively increase your distress tolerance by tuning into your body senses. Self-soothing, through your different senses, helps to reduce the intensity of any negative emotions you experience. Concentrate on your sense of hearing, sight, touch, taste, smell, and even add some movement.

Whenever you feel disturbed, overwhelmed, or in distress, focus on your primary senses. You can listen to the sounds of nature like the pitter-patter of raindrops, birds chirping, or even tune into the sound of traffic. You can listen to your favorite song or some soothing music. Another thing you can do is consciously shift your focus of vision onto something else. Count the number of colors you see in a specific room or focus on the texture of an object. You can even scroll through some of your favorite photos on your phone. Enjoying a small tasty treat is certainly pleasurable and gets you through a tough spot. It is not the same as emotional eating or binge eating. Ensure that it is only a small treat and not a full meal. Concentrate on the scents you notice in the air.

Try to identify some smells, or maybe you can break a smell down into different components. You can also place a couple of drops of your favorite essential oil on your hand or a cotton ball to calm yourself. Tune in to your sense of touch and notice how you feel when you play with a fidget toy, run your fingers over your desk, or concentrate on anything you are holding in your hand at the moment. Doing these things will effectively distract your mind from worrying about your distress, as you concentrate on something else. It is a diversion tactic that works brilliantly.

Technically, there are only five senses, but in DBT, even movement is considered as a sense. You can alter your emotional state by adding physical movement. It could be something as simple as dancing to your favorite song, walking, or anything that gets you moving.

While you are self-soothing using your senses make sure to concentrate only on one sense at a time, and it will teach you to be more mindful.

Chapter Nine: Mindfulness Tools for Fear, Insecurities, and Phobias

Dealing with Fear

Fear and anxiety can prevent you from moving forward in life. The longer you avoid them, the more difficult they become to deal with. Once you bring your awareness to different issues in your life that induce fear or anxiety, you stand a fighting chance to overcome them. In this section, let us look at specific mindful tips you can use to overcome any of your fears.

A Beginner's Attitude

Usually, we allow our past to prevent us from seeing things for what they are. Our past experiences prevent us from rationally observing situations. Instead of doing this, allow yourself to view things from a beginner's perspective. When you start thinking as a beginner does, you can see things for what they are because there is no other reality that exists. At times, experiences tend to cause anxiety or fear. For instance, if you had a string of bad relationships, you might be anxious to start dating again. Instead of doing this, view your next dating experience as something new altogether. Just because you didn't have a good relationship experience in the past, it doesn't mean your future experiences will be bad.

Don't Be Judgmental

We are often extremely judgmental, and most of us don't even realize it. Instead make an effort to be nonjudgmental. Notice whenever your mind starts claiming something to be either good or bad. Notice these thoughts, but don't react. Fear and anxiety often have certain messages they desperately want to convey. When you can calm yourself down and experience these feelings without any judgment, you get a better understanding of yourself. For instance, if you react strongly to a specific topic or an individual, ask yourself why you do this. By mindfully thinking about it, you give yourself a chance to understand the reasons for your behavior.

Be Patient

Unless you are patient, you cannot enjoy your present. Slow down for a while and be patient. Whenever you experience any anxiety or fear, observe your fear and carefully listen to it. It is not something you must shy away from. Ask yourself why you are scared. Are you afraid of success, failure, or judgment? Be a little patient and try to identify the cause for the fears. Learn to be mindful and live in the present while dealing with difficult emotions.

Accept Yourself

You must accept yourself truly, and unconditionally. Accept things for the way they are and accept yourself even before you change. Being honest and real with yourself is not an easy conversation, but it is quite important. You can stand in front of a mirror and ask yourself what is stopping you. Be compassionate and have a positive internal dialogue with yourself. If you don't like certain aspects of yourself, you can always work on improving them. You cannot do this unless you accept and embrace yourself for who you are.

Trust Yourself

The most common cause of anxiety includes self-created fears of success, failure, or judgment. This can make you feel hopeless and helpless. You don't have to feel any of these things, and you have the power to trust yourself. You can trust yourself that you will make mistakes, but you can feel proud for trying. You can trust yourself and find peace knowing that you will be fine, even if others don't agree with you. Regardless of what it is, you always have the power to believe in yourself.

Dealing with Insecurities

Insecurities and self-doubt can be crippling. If you keep second-guessing yourself and always doubt your abilities or the decisions you make, you cannot get ahead in life. Your insecurities and self-doubt make you incredibly vulnerable and scared of this vulnerability at the same time. Being overwhelmed due to this kind of thinking can effectively prevent you from taking the first step toward achieving your goals. If you spend all your time worrying about all the things that can go wrong, or your perceived flaws, you cannot enjoy life or any success that comes your way. In this section, let us look at some simple ways in which you can overcome insecurities and self-doubt.

Understand Your Expectations

Here is a simple question you must ask yourself: "What are my expectations, and are they realistic?" If you expect too much from your goal and expect it too soon, you are setting yourself up for disappointment. If your expectations are unrealistic, it will only worsen any insecurities you have. Everyone wants to be successful, and when you don't get the success you expect, disappointment creeps in.

Worst-Case Scenario

One question you must answer is: "What are you scared of? What is the worst possible outcome?" Once you answer this question, your fears will become obvious. Try to be as realistic as possible while determining the worst-case scenario. You probably realize that the worst-case scenario isn't as bad as it sounded in your head. Everything can be dealt with, and by preparing for the worst-case scenario, you can deal with it effectively if it does come true. Preparation is key to overcoming obstacles in life.

Mistakes Are Invaluable Lessons

Every mistake you make is not a failure. Instead, think of it as an opportunity to learn and improve. Mistakes are invaluable lessons that life is trying to teach you. So, learn from your mistakes, make any required changes, and move on. Instead of worrying too much about the mistakes you make, make a note of all the things you can do better the next time around.

Be Compassionate

Learn to be compassionate, not just toward others, but yourself too. Most of us are extremely critical of ourselves. If you think you are too hard on yourself, then give yourself a break. Everyone makes mistakes, and you're not an exception. If a loved one were to come to you for advice, wouldn't you be compassionate? Now, it is time to extend the same compassion toward yourself. You deserve some compassion. Stop being extremely critical of yourself all the time. Self-reflection and constructive criticism are important, but only to a certain extent. If you leave this self-criticism unchecked, it will turn into crippling insecurities.

Appreciation

Even if things don't go as you planned and you don't get the outcomes you desire, express your gratitude for all that you endured. You have come a long way, and it is not a journey to be taken lightly. Instead of unnecessarily focusing on unsuccessful attempts you might have made, concentrate on all the good you have in life. Instead of worrying about the criticism you have received from a couple of people, concentrate on the appreciation you have received from others. Be grateful for all of the good things you have in life, and you will feel better about yourself.

Visualize Your Success

A simple way to tackle insecurities and self-doubt is by visualizing the kind of success you desire. Visualize that you are confident, strong, happy; the true representation of all things you have always desired. Visualize what it would feel like to be successful. Make this visualization as clear and detailed as possible. For instance, if you want a promotion at work, visualize how you would feel if you got the promotion. Regardless of whatever your goal is, visualize that you have successfully achieved it, and it will give you the motivation to keep going.

Always Celebrate

Don't forget to celebrate yourself. Celebrate all the tasks you have successfully accomplished and completed. It could be going out for a meal with your loved ones, going on a holiday, or maybe even joining a hobby class. Acknowledge how far you have come in life. You have come a long way, and a lot of people don't even take those first steps, and well, look at you.

Dealing with Phobias

A phobia is an unreasonable fear to the extent that the person tries to avoid the particular object or situation that scares them. Merely thinking about the feared situation or object can make a person panic and become anxious. Whenever you are afraid of a specific situation, or an object, it is known as a specific phobia. There are various types of specific phobias, including fear of small spaces, fear of heights, fear of natural phenomena like storms, fear of deep water, fear of animals like spiders, fear of things like blood, needles, or pretty much anything else that you can think of. When it comes to phobias, the individual realizes the fear they experience is unreasonable or excessive in some way. For instance, it's normal to be scared of snakes, but someone with a phobia of snakes will avoid walking in parks because they are anxious about encountering snakes, even if it is unlikely. A phobia is also a type of anxiety or common mood disorder.

The physical symptoms of phobia include difficulty in breathing, trembling or shaking, feeling lightheaded, heart palpitations, and excessive sweating. Here are certain simple tips and techniques you can use to overcome any phobia you have.

Gradual Exposure

One of the most effective ways to desensitize yourself of any fear you have is by exposing yourself to the fear instead of avoiding it. This technique is known as gradual exposure. In this technique, you must start exposing yourself to your phobia and stop whenever the fear or anxiety becomes too much to bear. Every time you attempt this technique, try to push yourself a little farther and repeat until you no longer feel the panic associated with your phobia. This will allow you to control your fear instead of allowing the fear to control you.

Support Groups

There are plenty of support groups you can join to meet other like-minded people. Doing this will enable you to understand you are not alone, and there are others like you. Talking to others might also help relieve any anxiety you feel. It could also provide insights into the different tips that others follow to overcome their phobias, which might come in handy.

Gather Information

Most phobias are based on irrational fears, and if you gather information about what scares you, it becomes easier to understand your phobia. For instance, if you are scared of flying, try to understand how planes work and all the safety measures they have in place. By gathering information, you will be presented with hard facts that help relieve the irrational fear you experience. It might not help you to overcome the phobia entirely, but once you understand more about it, you can reduce the fear you experience.

Flooded Exposure

In this technique, you will be completely exposed to your phobia. You are required to endure your fear until it has run its course. By repeating this exercise several times and facing the fear until you realize the situation is not harmful, you can tackle your phobia. In the end, you will realize your fears might be unpleasant, but they are not life-threatening. It helps to ease your mind and put your fears to rest.

Fear Ladder

The fear ladder technique is quite similar to gradual exposure. In this technique, you start with a very simple situation and slowly work your way up until you are face-to-face with your phobia. For instance, if you are scared of snakes, you can start by looking at a photo of a snake, then looking at a snake through a window, and maybe even standing next to it in a safe environment (like a petting zoo). This might not be an ideal technique to deal with all types of phobias, but it is an effective method to deal with any phobia associated with objects or things.

All the different techniques to overcome phobias discussed in this section will help regulate your fear instead of allowing your fear to regulate you.

Chapter Ten: Mindfulness Meditation Techniques for Anxious Minds

When you start experiencing too much stress, it triggers anxiety. Anxiety triggers the fight or flight response in your body, which makes you feel like you are always on alert. When this feeling does not go away and instead becomes background noise, it's time to seek help. The three simple ways to calm your anxiety are: mindfulness techniques, concentrating on your breath, and focusing on your body.

Mindfulness enables you to deal with difficult feelings without suppressing, overanalyzing, or encouraging them. When you start to feel and acknowledge all your emotions without any judgment, it becomes easier to cope with them. Mindfulness gives you a chance to explore the underlying reasons for the stress or anxiety you feel. Instead of wasting your energy trying to ignore or fight your anxiety, mindfulness allows you to understand the reasons behind it. Mindfulness also creates a safe space around you so that the feelings don't overwhelm you. When you can understand the underlying causes of your stress, you can take corrective actions to prevent such situations in the future.

Meditation is one of the best techniques you can use to calm an anxious mind. Meditation is also the key to mindfulness. When you start becoming aware of the moment you live in, you get a chance to explore different inner resources you never knew you had. Also, this approach effectively calms your mind. We usually have a good idea of all the things that we want or don't want in our lives. However, we can get overwhelmed by the different situations that we face and forget about this basic awareness we have. Meditation enables you to access this awareness for improving yourself.

Meditation Techniques

In this section, let us look at two simple meditations you can follow to get rid of anxiety.

Meditation for Relieving Anxiety

By following this exercise regularly, you will certainly feel more relaxed and will be in a better position to deal with any anxiety you experience. This simple meditative exercise helps you to let go of tension, alleviate stress, and enjoy a calm state of being.

Start by finding a calm and comfortable spot for yourself. You have the option of either sitting cross-legged on the floor or lying down. Now, place your arms by your side and keep your legs stretched out straight. Concentrate on reining in your focus; you can either close your eyes or focus on a single spot or object. Keep your eyes closed and breathe in deeply through your nose and allow the oxygen to fill up your lungs. Gently breathe out and allow the air to slowly leave your body.

Inhale slowly and deeply through your nose. Exhale slowly through your mouth.

Inhale slowly and exhale slowly.

Inhale slowly and exhale slowly.

Keep doing this and your body will start to feel calmer and more relaxed. You don't have to do anything right now, and the only place you are required to stay is right here, in the moment. You deserve this kind of calmness and it will improve your overall productivity and focus. You have nothing to worry about, and you need this time for self-relaxation.

Keep breathing slowly and maintain a steady rhythm of breathing. While doing this, slowly shift your attention to your body. How does your body feel? How do you feel in different parts of your body? Do you notice any tension or stress in a specific area? You don't have to try to change anything, and all you need to do is merely observe. Don't be judgmental of the feelings and tell yourself that your feelings are valid and justified. Notice any signs of stress or tension, and make a note of it, so that you can get back to it later.

Start performing a mental scan of your body; starting from the crown of your head to the tip of your toes. Take stock of every area and notice any tension present. Observe how different parts of your body feel. Slowly start with the crown of your head, move onto your neck and shoulder area, then down to your chest, arms, tummy, hips, legs, and do this until you reach the tip of your toes. Now, go back to your previous observations and try to notice which area seems the tensest. Start concentrating on that area and allow your muscles to relax.

Any tension you notice is caused because of the involuntary contraction of muscles. By voluntarily relaxing them, you are willing them to let go of any stress. Now, notice how it feels to be relaxed. Imagine this warm and wonderful sensation of relaxation move through the length of your body and spread all over you. As you feel your body relax physically, the mental anxiety you feel will decrease. Whenever you breathe in, your body is absorbing oxygen and relaxing, and while you exhale, it is letting go of anxiety and carbon dioxide. Visualize a small ball forming in front of your body whenever you exhale. The ball contains all the anxiety and tension present within your body. With every breath you expel, this ball starts getting bigger. Imagine the different areas of tension and visualize that you are expelling the anxiety present within.

Now, it is time to perform a body scan to notice how you feel. Do you feel lighter and a little calmer? Visualize that your body is made of a solid substance that can melt. Right now, your body might feel like it's made of an opaque and strong solid substance. Visualize that a feeling of warmth starts slowly spreading from your hands and feet throughout your body. This warm energy is slowly melting away the solid substance and making your body more fluid. As your body starts becoming a little soft, you start feeling calmer and more relaxed. Revel in this feeling of relaxation and concentrate on holding onto it.

It is time to concentrate on your thoughts. You can use a mantra, positive affirmations, or even a phrase to calm your mind. You can repeat the word "Relax" over and over again to calm your mind consciously. Even if there are various thoughts present in your mind, don't dwell on them right now. You can get back to them after the meditation ends. For now, concentrate only on completely relaxing your mind. By repeating "Relax" over and over again, this word gets embedded into your subconscious and will stay with you all day long.

Breathe in and say the word "Relax."

Breathe out and say the word "Relax."

Keep doing this until you finally feel calm and composed. Whenever your mind starts to stray away, concentrate on this word once again. This feeling of calm you are experiencing right now will stay with you all day long. You can tap into its energy whenever you feel anxious or stressed.

Once you are ready, it is time to end the meditation. To do this, shift all your attention to your breathing again, and concentrate only on your breath. Do this for a minute or two, then slowly open your eyes and get back to your routine.

Meditation for Instant Relief

The most common symptoms of anxiety include tension in muscles, rapid and shallow breathing, unsettling thoughts, and involuntary tightening of muscles in the body. By using this simple meditation, you can obtain quick relaxation whenever you feel anxious.

Start by finding a calm and comfortable spot for yourself. You can either sit or lie down according to your convenience. Start concentrating on your breathing and take in deep and long breaths. Breathe in slowly through your nose and exhale slowly through your mouth. Maintain a steady and calm rhythm and continue breathing in slowly and deeply. By following this simple breathing technique, you can calm your mind and replenish your body with plenty of oxygen. While doing this, try to make yourself as comfortable as possible. In fact, your comfort must be your only priority right now. If you are feeling anxious, here are some simple phrases you can use to calm yourself down.

"I am feeling anxious, but I know I am fine. This feeling will pass, and I don't have to worry. I am in a safe place, and no one can harm me. I know I am safe, but my anxiety scares me. I know I will be fine in a while. I will wait for the anxiety to reduce, and I will concentrate on making myself comfortable. I can control my thoughts, and I can relax my mind."

You can either say this out loud or repeat it mentally until you feel calm. Shift all your attention to the sentences you utter and don't think about anything else. Even if random thoughts start popping into your mind, let them pass through. Don't judge them, and certainly don't try changing them. Focus only on what you desire to attain from this meditation exercise.

Keep reassuring yourself and calm your mind. Breathe in slowly and deeply. Breathe out slowly. Keep repeating these messages until your mind is calm, so that your body can escape any apparent danger that it is responding to. Since your mind cannot distinguish between real or imagined danger, this reaction is triggered by any form of stress. Your body also releases plenty of adrenaline when you are stressed, and it might result in trembling. To get rid of this, you can physically shake yourself out.

Start by shaking your hands as if you were shaking off water. Allow your hands to go limp and place them by your side. After your hands, shake your arms. Then shake your shoulders, neck, head, legs, and then your entire body. You are essentially wiggling your body to wiggle out the anxiety you feel. Do this for a couple of minutes, and you will feel better.

Now, place your hands by your side and feel the relaxation course through your body. Keep breathing evenly and count to ten. Think all sorts of calming thoughts. Here are the steps you can follow while breathing to the count of ten:

One — I am calm.
Two — I am relaxed.
Three —   I am calm.
Four —I am relaxed.
Five — I am calm.
Six —   I am relaxed.
Seven —   I am calm.
Eight —I am relaxed.
Nine —I am calm.
Ten — I am relaxed.

Now, concentrate all of your attention on your body and look for any areas where the muscles feel tight or tense. Whenever you notice such tension, try to relax your muscles consciously. Start by letting your jaw go slack so that your teeth don't touch. After this, lower your shoulders and move them gently. If you want, you can even swing your arms and allow them to go limp. Place your arms over your head and stretch as high as you can. Feel the stretch in your muscles, and slowly turn your head toward the left and then right. Look right ahead and then down. By doing this, you are removing any traces of anxiety left behind in your body.

Keep at it until you feel relaxed. Once your mind feels calm, it is time to end the meditation. Concentrate on your breathing for a minute or two and then get back to your reality.

Chapter Eleven: OCD – 11 Mindful Ways to Beat the Obsessive Mind

What is OCD?

OCD or obsessive-compulsive disorder is a mental health issue where an individual experiences unwanted sensations and thoughts repeatedly, or they have a compulsive urge to do a specific activity over and over. Some people tend to experience both obsession and compulsion. Biting your nails repeatedly or thinking undesirable thoughts is not OCD. An example of an obsessive thought is to believe that specific colors or numbers are good or bad. An example of a compulsive habit would be to wash your hands five times whenever you touch something you believe is dirty. You might want to stop doing these things, or thinking these thoughts, but you feel powerless to stop yourself. We all tend to have certain thoughts or habits we keep repeating at times. However, those with OCD indulge in actions or thoughts that take up at least one hour daily, are not enjoyable to them, are absolutely beyond their control, and negatively affect their social life, work, or any other aspects of life.

Types of OCD

There are various forms of OCD and they can usually be categorized into the following four categories:

Contamination

In this type of OCD, you might experience a compulsive need to keep things clean. It is almost as if you are terrified of things that might be dirty. Contamination isn't restricted to physical things but can be mental, too—feeling like you are being treated like garbage or dirt.

Rumination

An undeniable obsession with thinking about a specific line of thought. These thoughts are often intrusive and can be disturbing or violent at times.

Checking

The constant urge to check and recheck things like alarms, locks, doors, ovens, switches, and so on. It could also mean you cannot stop yourself from thinking you might have a mental condition like schizophrenia or that you are pregnant, for example.

Order and Symmetry

An obsession with order or symmetry and the need to have things arranged in a certain way.

Most people with OCD usually realize that their habits or thoughts make no sense. They aren't compelled to have certain thoughts or perform certain actions because they find them pleasurable, they do it because they cannot stop themselves. Even if they do stop, they feel so terrible about it that they resume their thoughts and habits. Here are some examples of obsessive thoughts and behaviors:

• Any worry about getting hurt or others getting hurt.

• Continuous awareness of breathing, blinking, or any other body sensations.

• Suspecting or doubting others, even when you have no reason to believe your suspicions.

• The constant need to count things like coins, bottles, or steps.

- Following a specific order while doing certain tasks every time, or doing them a couple of times because you believe it is good.

- An extreme fear of using public restrooms, touching doorknobs, or shaking hands.

Risk Factors

The causes of OCD aren't entirely clear, but it is believed that stress often worsens the symptoms. Signs of OCD usually start appearing in teens or young adults. A couple of risk factors for OCD include having a parent, child, or sibling with OCD, experiencing trauma, dealing with physical or sexual abuse in childhood, depression or anxiety, or any physical differences in specific areas of the brain. Trained therapists and medical professionals can diagnose OCD.

Mindful Ways to Overcome OCD

Mindfulness is an effective way to tackle anxiety, since it places great emphasis on understanding and accepting one's thoughts. Whenever a disturbing thought pops into your head, you allow it to stay in your mind without attaching any significance to it. You don't try to judge the thought, you don't change it, and then it eventually goes away. Mindfulness teaches you to let the thoughts pass through, instead of spending your time thinking about whether they should or should not exist in the first place. This skill certainly comes in handy while dealing with intrusive thoughts associated with OCD.

Technique #1: No Comparison

Never compare yourself with those around you. Just because someone seems happy or relaxed, don't feel that you need to measure up. Don't worry about societal standards and expectations. Don't feel guilty for any obsessions or compulsions you have. Understand that OCD is a disorder, and you did not ask for it. Learning to manage it is your responsibility.

Technique #2: Don't Be Alone

At times, you might feel an urge to withdraw or indulge in any negative thinking patterns. Often the intrusive thoughts prompted by OCD are directed toward self-harm or any harmful behavior. If you feel like you are leaning toward something undesirable or are thinking worrisome thoughts, don't be alone. Get out and spend some time with your loved ones or seek professional help. Practice simple mindfulness skills discussed in the previous chapters in order to become more aware of your thoughts.

Technique #3: Worry Hour

It becomes difficult to stop yourself from obsessing over certain thought patterns or exhibiting compulsive behaviors when you have OCD. Instead of trying to ignore or avoid such patterns, you can set aside a specific time slot for thinking. It could be your worry time. During this period, allow your mind to run free and allow all the different thoughts to pass through your mind. Don't change the thoughts, and don't judge yourself for having them.

Technique #4: Listening to Music

A simple distraction technique that works wonderfully is listening to music. Whenever you feel your OCD kicking in, take time out and listen to some music you enjoy. It could be something peppy, upbeat, or even relaxing. Listen to your favorite songs, and if you can, try singing along. Do this even if you aren't a good singer. Just go with the flow and allow your mind to unwind.

Technique #5: All-or-Nothing Attitude

Let go of the all or nothing mentality while dealing with OCD. There will be times when you slip up and fall back into your old patterns of OCD. If that happens, tell yourself it is just a slip-up and move on. Don't start obsessing over it, or you might end up forming a new compulsion or obsession. Mindfulness doesn't come easily, and it takes plenty of practice. When you are in it for the long haul, there is the chance you will end up making mistakes. As with any other skill, unless you keep practicing it, you cannot master mindfulness. In this process, there will be a couple of slip-ups. So, expect the slip-ups and don't allow them to harm your morale.

Technique #6: Expecting the Unexpected

An obsessive or intrusive thought will not give a warning before it presents itself. You can have such thoughts at any place and at any time. At times, an old thought might present itself, or a new one might develop. Regardless of what it is, don't allow yourself to be caught off guard and always prepare yourself for the unexpected.

Whenever an obsessive thought pops into your mind, use any of the previous techniques to deal with it.

Technique #7: Don't Avoid Thinking

Don't waste any of your time and energy trying to avoid your thoughts or prevent yourself from thinking. It is impossible to do this, and the results often produced are the opposite of what you expect. In fact, the more you tell yourself you're not supposed to think about something, the stronger you will get the urge to think about it. If you want to stop certain thoughts, you merely need to allow them to pass through your mind. When you don't give them any importance or attach any significance to them, they will slowly disappear.

Technique #8: Acknowledge Your Thoughts

Don't try to stop any of your thoughts. This piece of advice is perhaps easier said than done. Keep in mind that thoughts are nothing but a collection of words that randomly pop into your mind and cannot be dangerous unless you act on them. You don't have to take your thoughts seriously, just because they appeared in your mind. You are under no obligation to act on them. Don't judge yourself based on your thoughts. It might be difficult, but don't push your thoughts away; try to acknowledge them. When you try to push them away, they will come back to haunt you and you might increase your obsession with them. To avoid all this, it is best to acknowledge that they exist, rather than to resist them. Allow your thoughts to come and go without any judgment. Don't get upset with yourself for having these thoughts. As mentioned earlier, they are nothing more than a random set of words. Understand that your thoughts are real, and they don't reflect poorly on you. Unless you act on them, thoughts have no power.

Technique #9: No Toxicity

The symptoms of OCD worsen with an increase in stress. By managing and limiting the stress you experience, you can reduce the intensity of the OCD symptoms. A great way to let go of unnecessary stress in your life is by getting rid of all toxic people. People who radiate negativity will bring you down mentally. Instead, concentrate on surrounding yourself with your well-wishers and all those who genuinely want to help you. Start prioritizing all the relationships you have in life. Try to replace the bad ones with more desirable and positive ones.

Technique #10: Anxiety is Not the Same

The intensity of anxiety you experience will differ. At times, it can be quite intense, and at times, it can be mild. Anxiety is almost like the ocean. On some days, the waters are rough; on other days, the ocean is still, and it's smooth sailing all the way. So, prepare yourself to deal with your symptoms. There will be good and bad days. It is important to relish the good that comes your way, but prepare yourself for the sour patches too.

Technique #11: Mindfulness Meditation

Mindfulness meditation is a great way to regather your thoughts and calm your mind. It enables you to start viewing your thoughts objectively without giving in to the urge to judge or over-analyze them. You start to become more aware of all your thoughts. It also teaches you to detach yourself from your thoughts and view them at arm's length. The likelihood of being affected by troubling thoughts, including obsessions that are associated with OCD, can be reduced by following this strategy. To start mindfulness meditation, concentrate on taking a couple of deep breaths. While you breathe, notice the different thoughts, fears, sensations, worries, or anxieties going on in your mind. You merely need to notice them and don't try to push them away. Try to observe what happens when you leave the thoughts alone and allow them to pass through your mind freely. While doing this, you might feel more anxious during the initial stages. It happens because you are finally face-to-face with all the different thoughts, worries, and anxieties that trouble you. After a while, you start getting accustomed to them and can allow them to sit in your mind without acting upon them.

You can use any of the different mindfulness meditation techniques discussed in the previous chapters.

Chapter Twelve: How to Stop a Panic Attack with Mindfulness

What is a Panic Attack?

The abrupt onset of potent discomfort or fear that peaks within a few minutes; a panic attack includes at least four of the symptoms discussed below:

- Discomfort or chest pain.
- Increased heart rate, palpitations, or pounding of the heart.
- Excessive sweating.
- Shakiness or trembles.
- Feeling lightheaded, faint, dizzy, or even unsteady.
- Cold or hot flashes.
- A feeling of numbness or tingling sensations.
- Derealization.
- Depersonalization.
- Abdominal discomfort or nausea.
- Feel like you are being choked.
- Shortness of breath.
- Fear of going crazy or losing absolute control.
- Fear of death.

Physical symptoms like heart palpitations or a pit forming in your stomach are commonly associated with anxiety. Therefore, it might sound quite similar to a panic attack. The one thing that differentiates anxiety from a panic attack is that the intensity of the symptoms and their duration is longer in the latter. Panic attacks take about ten minutes to intensify fully, and then they start subsiding. Considering the intensity of the symptoms associated with it, they resemble the symptoms associated with breathing disorders, cardiovascular diseases, or any other chronic illnesses. Usually, individuals with panic disorders often seek immediate medical help, because they wrongly believe they are suffering from a life-threatening issue.

Panic attacks occur out of the blue and can occur even when you are calm. Panic attacks are commonly associated with panic disorder, but individuals with other types of psychological disorders can also experience them. For instance, for someone who has a social anxiety disorder, they might have a panic attack before giving a speech, or someone with OCD might have a panic attack when they are stopped from engaging in any compulsive rituals.

Panic attacks are not only frightening, but they are extremely distressing as well. Due to this, all those who experience frequent panic attacks worry about having another attack and try to make lifestyle changes to avoid them. For instance, they might start avoiding certain places that trigger panic or start exercising more to stabilize their heart rate.

Tips to Tackle Panic Attacks

Recognize the Symptoms

Go through the list of symptoms discussed in the previous section. Spend some time on self-reflection and think of all the instances in the past when you experienced a panic attack. What were the different symptoms you noticed while having a panic attack? You would identify with at least a few of the symptoms discussed in the list mentioned above. So, the next time you start experiencing any of these symptoms, it is a sign that you are about to have a panic attack. At times, merely recognizing that you are having a panic attack reduces the stress associated with the occurrence. For instance, when you recognize you are having a panic attack and not a heart attack, it becomes easier to reassure yourself that it will pass, and you don't have to worry about it. By taking away the fear that something bad is happening to you, you can calm your mind. It also gives you a chance to practice the other techniques discussed below to reduce the intensity of the panic attack.

Deep Breathing

Hyperventilating is one of the most common symptoms of a panic attack. Hyperventilating can also increase the fear associated with experiencing such an attack. To counteract these feelings, use deep breathing. Once you get your breathing under control, the likelihood of hyperventilating is also reduced. Since hyperventilation worsens the symptoms of a panic attack, deep breathing comes in handy to help control hyperventilation.

Start concentrating on taking in long and deep breaths. Breathe in slowly and breathe out slowly. Always inhale through your nose and exhale through your mouth. Notice how you feel when the air starts to slowly fill your chest cavity along with your belly. Then permit yourself to exhale slowly through your mouth. Here's a simple breathing exercise; you can try to reduce the chances of hyperventilation.

Inhale to the count of four—one, two, three, four.

Hold your breath for a second.

Exhale to the count of four—one, two, three, four.

Repeat this exercise until your breathing normalizes.

Practicing Mindfulness

Panic attacks often make individuals feel like they are losing grip of reality, or they are detached from it. To counteract such feelings, you can use mindfulness to keep yourself grounded. Instead of the panic that starts to engulf your body, concentrate on all the physical sensations that seem familiar to you. For instance, you can dig your feet into the ground, or feel the texture of the surface under your feet. You can run your hands over your trousers and feel the soft material. These are certain specific sensations that enable you to stay in the moment, instead of getting overwhelmed by unpleasant emotions or feelings. Any form of sensory stimulation helps divert your mind and reduces the symptoms of a panic attack.

Shut Your Eyes

Certain triggers can quickly overwhelm you. For instance, loud music or even violence on television can trigger a panic attack. When you are in an environment with plenty of stimuli, it increases the stress on your senses and worsens a panic attack. The simplest way to shield yourself from excessive stimulation is by shutting your eyes. Whenever you feel the panic creeping up on you, shut your eyes as tight as you possibly can, and take a step back. By blocking all the unnecessary stimuli, it becomes easier to concentrate on your breathing. Once you start concentrating on your breathing, it becomes easier to regulate your emotions. It is a simple technique of mindfulness that allows you to stay grounded in the moment.

Muscle Relaxation

Forget about everything for a while and turn all your attention toward your body. Concentrate on every part of your body and try to understand what you feel. Do you think your muscles are tightening? Can you feel any sensations in your toes or hands? What do you feel when you wiggle your toes? Do you notice any changes in the overall sensations as you start concentrating on your breath? How does your body feel whenever you inhale and exhale? Regardless of what you feel, allow yourself to feel it fully. Don't try to change any of your feelings and merely understand what you feel.

Focus

Another simple technique you can try is to find a focus object and shift all your attention toward it during a panic attack. Zero in on a specific object that's in your line of vision and shift all your attention to it. The object must be the focus of your mind. Try to notice everything possible about it consciously. For instance, you might notice how a clock's hands jerk as time ticks away. Perhaps the clock is crooked, or it isn't aligned properly. Notice all these details and try to describe the colors, patterns, shapes, and sizes to yourself. Describe everything you can possibly think of about the object and do this until you feel calmer. By giving your mind something else to concentrate on, you distract it from the panic creeping up on you.

Light Exercise

When you start panicking, your heart starts beating faster, and it increases the blood flow in your body. The simplest way to redirect this blood flow toward something more positive is by engaging in some light exercise. When your blood starts pumping rapidly because of exercise, your body starts releasing endorphins. Allow your body to be flooded with endorphins—it helps to improve your mood. Whenever you start feeling stressed, opt for gentle exercises like swimming or walking. However, don't attempt any form of exercise if you start hyperventilating or are struggling to breathe. The first thing you must do in such a situation is to catch your breath. If you are standing, and start hyperventilating, sit down, and place your head between your knees and try to take slow and steady breaths. Think about a calming place or repeat a positive affirmation to calm your mind and steady your breathing. Once your breathing is stabilized, you can go for a short walk and breathe in some fresh air.

Happy Place

We all tend to have certain places that make us insanely happy. Whenever you have a panic attack, think about your happy place. What is the one place in this world that always makes you happy and instantly lifts your spirits? Perhaps it is the seashore, the comfort of your bed, or a cabin in the mountains. Once you are aware of the happy place, start visualizing it as if you are there. Make your visualization as detailed as possible and think about all the different sensory experiences associated with your happy place. For instance, if it is a seashore, visualize a sandy beach with a crystal blue sea and bright sunshine. Visualize how you would feel when the sunlight falls gently on your face, while the water caresses your feet, and your hair is blowing in the sea breeze. Revel in all these sensory experiences and visualize them. It helps calm your panicking mind.

Repeat a Mantra

You can either repeat a mantra or come up with a positive affirmation. By repeating the same sentence a couple of times, it effectively distracts your mind and makes you feel calm. Since a panic attack can last for ten minutes or more, you must keep calm during this process. To help stay calm, use a mantra or a positive affirmation like, "I am panicking, but it will pass," or anything else that works for you. Keep repeating it until you feel the panic calm, and then finally disappear.

Lavender

The scent of lavender is commonly used to relieve stress. Lavender can help your body relax, especially when you are prone to panic attacks. So, always keep a little lavender essential oil at hand. Whenever you experience a panic attack, simply dab a couple of drops of this oil on your hand or forearms and smell it. Try to breathe in its calming scent. Alternatively, you can also drink some lavender or chamomile tea. Both these ingredients are extremely calming and soothing for your mind and body. If you drink a cup of chamomile tea before going to bed, it can improve the quality of sleep you get at night.

Acknowledge Your Surroundings

For some people, it helps when they close their eyes while experiencing a panic attack. However, for others, it only worsens the panic they feel. If you are among the latter, then keep your eyes open. Whenever a panic attack occurs, keep your eyes wide open and try to acknowledge your surroundings. It essentially enables your mind to concentrate on the present instead of worrying about the overwhelming panic you feel. Start by looking at your hands, your feet, the ceiling, the ground under your feet, or anything else around you. It might sound a little silly, but it works. Start by mentally acknowledging everything that you see around you. Concentrate on one object and then move onto another. It is also a great way of teaching yourself to be mindful of the present without getting overwhelmed by the panic attack.

By following the simple tips discussed in this section, you can effectively manage a panic attack without allowing it to control your life. Also, all these tips will make you feel more mindful in your usual life. Whenever you feel anxious, you can try using the steps to reduce your anxiety to prevent a panic attack.

Chapter Thirteen: Trauma and PTSD – How DBT and Mindfulness Can Help

What is PTSD?

Anyone who has endured a trauma can find themselves experiencing all the emotional challenges associated with the traumatic event, long after its occurrence. People tend to experience psychological difficulties after any trauma they have lived through, but the intensity of their symptoms often reduces with time. However, this doesn't happen in individuals with post-traumatic stress disorder (PTSD). In this disorder, an individual continues to experience the distress without any signs of it letting up anytime soon.

The *Diagnostic and Statistical Manual of Mental Disorders (DSM-5)* is a manual used by clinical professionals for diagnosing mental health issues. Initially, PTSD was classified as an anxiety disorder by this manual, but now it has been reclassified into the category that includes disorders associated with trauma and stressors. PTSD is often developed after experiencing a traumatic event. This event could be an isolated one, or it could be in the form of recurring and chronic traumatic experiences. There are a variety of emotional difficulties and symptoms associated with PTSD that induce significant distress and impair an individual's ability to interact socially, work effectively, and other important areas of life.

Some of the factors that can contribute to PTSD include: the type of trauma and its intensity; a person's gender, marital status, age, physical health condition, and mental health condition; emotional responses during a trauma; emotional support system; and experiencing additional stressors after the occurrence of trauma. There are different types of PTSD. Three subtypes of PTSD include: preschool, complex, and delayed onset PTSD. If you are suffering from PTSD you should seek professional advice and support.

It is not just adults; even young children can experience symptoms of PTSD. Whenever they live through any traumatic events or witness traumatic events, they can experience emotional distress after the event occurs. The symptoms of derealization or depersonalization are considered as a dissociative subtype of PTSD. Derealization is the term used to describe a condition where a person senses things around them as if they aren't real, and they feel unfamiliar with or disconnected from the world around them. In depersonalization, the individual starts experiencing events as if they are observing from outside their body and they themselves aren't real. In delayed onset PTSD, an individual doesn't start experiencing the different symptoms until at least six months after of the occurrence of a traumatic event. At times, individuals can experience certain isolated and severe instances of trauma, like mugging, sexual assault, or a horrific accident. Such instances are believed to be isolated since the chances of their recurrence are not high. There are different types of trauma, which can be recurring, like domestic violence, childhood neglect, or sexual abuse. Whenever a person has complex PTSD, they start reliving the event over and over, long after it has transpired.

An individual's experience of PTSD will be unique to them. The common symptoms of PTSD are re-experiencing or reliving aspects of what happened, hyperarousal, avoidance, and negative thoughts and beliefs. Let us look in more detail at these symptoms:

• Thinking any upsetting thoughts or memories associated with a traumatic event frequently or constantly.

• Recurrent nightmares and an inability to fall asleep peacefully at night.

- Extreme feelings of distress whenever you are reminded of the traumatic event.
- Experiencing physical symptoms like palpitations or excessive sweating whenever you remember the traumatic event.
- Reliving the event through flashbacks where you feel as if the traumatic event is occurring all over again.
- Signs of hyperarousal include difficulty falling asleep or staying asleep; being susceptible to angry outbursts; feeling always on guard or that there is often danger around the corner; startling easily; feeling jumpy; and having trouble concentrating.
- If you make a conscious effort to avoid having any conversations about the traumatic events or avoid the thoughts of feelings associated with those events, it is a sign of avoidance. If you actively avoid all the places and people who remind you of the traumatic event or keep yourself engaged in avoiding thinking about the trauma this is also a sign of avoidance.
- The common signs of negative thoughts and beliefs include trouble remembering specific aspects of the traumatic event; losing interest in activities you once enjoyed; feeling like an outsider; having trouble experiencing positive emotions; or feeling a strong urge to withdraw from others.

Most of these symptoms are your body's natural defense mechanism to the stress it endured during the traumatic event. During the traumatic event, your body switches on its fight or flight response—a natural response to any situation that is dangerous or viewed as a threat. The same mechanism is triggered whenever you are reminded of the trauma you endured. By understanding your body's natural response, you'll be better equipped to deal with any of the symptoms associated with PTSD.

Using DBT for PTSD

Those with PTSD often struggle to effectively and constructively manage their emotions, the way people with BPD struggle. It is one of the reasons why DBT is used for treating PTSD. If you have PTSD, then you might have trouble forming and maintaining relationships, and you might also experience an urge to indulge in self-destructive actions, like deliberately harming yourself.

Researchers from the Central Institute of Mental Health in Mannheim, Germany, conducted a study to explore the effectiveness of using DBT to treat PTSD. Researchers studied the impact of an intensive treatment combining DBT and CBT for treating PTSD in women who suffered sexual abuse in their childhood. They referred to this treatment as DBT-PTSD. After three months, the researchers noticed that this joint approach helped significantly reduce PTSD symptoms like anxiety and depression. The symptoms displayed by the participants were still reducing six weeks after treatment, with the women continuing to use skills they had learned to treat the PTSD. However, all the research about DBT-PTSD is still in its infancy, and there is a need for plenty of research. The preliminary findings certainly suggest that DBT is a great way to treat PTSD, and it offers a promising scope.

Mindfulness and PTSD

Mindfulness can be used for dealing with symptoms associated with PTSD. The concept of mindfulness has been around for centuries, and mental health professionals are only now able to understand the various benefits associated with it. Individuals with PTSD find it exceedingly difficult to distance themselves from all the unpleasant memories, emotions, and thoughts associated with the trauma they endured. They can feel extremely preoccupied with or distracted by all these thoughts. Due to this, individuals with PTSD have a tough time focusing on things that matter in life, such as personal or professional relationships. They cannot enjoy activities that they used to enjoy before the trauma occurred. Mindfulness is a great technique that enables such people to get back in touch with their reality and learn to live life in the present. It also reduces the intensity of the unpleasant emotions or memories they experience due to PTSD.

There isn't sufficient research about the association between mindfulness and PTSD. Mental health professionals have only recently started to understand the benefits offered by therapy like mindfulness. However, all the research conducted so far does point out that mindfulness can significantly reduce any anxiety an individual feels. Therefore, it is safe to say that mindfulness is an effective way to reduce stress and anxiety associated with PTSD.

In this section, let us look at different mindfulness skills you can start practicing to manage and reduce the intensity of PTSD symptoms.

Being Aware

One of the fundamental skills of mindfulness is awareness. It is about your ability to concentrate only on one thing at any given point in time. Awareness not only means being aware of all things going on around you—you also need to be able to recognize them and become aware of all things going on within you, like your feelings and thoughts.

Another aspect of awareness is the ability to live in the moment, without indulging in unhelpful rumination or worry. If your life is guided by all emotions, thoughts, and feelings associated with a trauma, it becomes exceedingly difficult to notice and take part in life as it passes you by. Instead of living life on autopilot, which is mostly guided by your past, it is desirable to live in the moment.

No Judgments

Another important aspect of mindfulness is evaluating your experience without any judgment. It is known as non-evaluative observation. It merely means you must be able to look at things objectively, without categorizing them as good or bad. Non-judgmental observation is important for practicing self-compassion. Unless you can be compassionate toward yourself while evaluating your experiences, you cannot change your thought patterns. A major factor responsible for negative thinking is self-criticism. When you are compassionate to yourself while reviewing any traumatic incident, you can begin to let go of any painful memories or thoughts associated with it.

Open to New Experiences

Living with PTSD is often tricky. All the trauma can prevent you from exploring any new possibilities present in life. At times, it can be difficult to view things as they are instead of allowing any preconceived notions or biases to guide your judgment. For instance, if you believe that there is nothing good left in your life, then even when brilliant opportunities present themselves right in front of your eyes, you cannot see them. Or maybe when you try to change yourself, you believe that you cannot change, so it becomes exceedingly difficult. If you want to let go of the different symptoms associated with PTSD, it is time to open-up yourself to new possibilities and all the wonders of life.

Getting Out of Your Head

Unknowingly, most people with PTSD often get caught up in anxiety or worries and get stuck in their own heads, so to speak. Here's a simple exercise you can try to become more mindful and increase your awareness.

Find a calm spot and make yourself comfortable. For the duration of this exercise, get rid of all distractions and place your phone on silent. You can either lie down on your back or sit in a chair. While sitting down, ensure that you keep your back straight, shoulders relaxed, and your arms by your side. Simply close your eyes and shift all your attention to your breathing. Notice how your breath feels as it enters and exits your body. You can concentrate on the physical aspects of breathing—the way your abdomen rises and falls as you inhale and exhale. Additionally, you can place one hand over your stomach to notice this movement, and it enables you to stay grounded.

Allow yourself to get lost in this experience and focus only on your breathing. If your mind wanders, gently guide it back to the rise and fall of your lower abdomen. Visualize that you are riding the gentle waves of your breath. Also, notice how it feels whenever your mind wanders, it gives you an idea of the signs to look for before you give in to any distractions. Keep doing this for as long as you want or until you feel calm. Before you try this exercise, practice the technique of mindful breathing. We seldom pay attention to the way we breathe.

Since breathing is an involuntary function, it doesn't require conscious thought. Mindful awareness means you must concentrate on your breath to make it even, rhythmic, and prevent it from becoming shallow or rapid. Follow the mindfulness exercise mentioned above at least once daily to improve your awareness. Over time, it will give you better control over your thought patterns and emotions.

The first time you try this exercise, your mind might wander several times. Mindfulness is like learning to drive. You cannot master it until you practice. Don't ever get discouraged if your mind starts wandering during the exercise. It is quite normal. Whenever you notice any thoughts popping into your mind, make a note of them, and think about them after the exercise. Also, don't judge any of your thoughts and merely observe.

Try to practice the different mindfulness skills discussed in this chapter as often as you can, while you go through your daily life. With practice, mindfulness will come to you naturally, and you will become more aware of all your life experiences. That, in turn, will make it easier to cope with symptoms of PTSD.

Chapter Fourteen: Relapse Prevention

Lapse and Relapse

Recovering from and dealing with any mental health issue is rarely straightforward. The path toward relief from symptoms is seldom steady, and a realistic scenario often includes some setbacks. These setbacks can be in the form of lapses and relapses.

The terms lapse and relapse might sound a little confusing and are often used synonymously, but they are quite different from one another. Whenever a clinician describes a lapse, they are referring to a condition that is quite normal. Lapse usually means a brief return to feeling low or engaging in any undesirable thinking or behaviors. It usually is a temporary situation and is quite common. On the other hand, relapses are slightly more complicated and trickier to deal with. A lapse can quickly transform itself into a relapse if the symptoms associated with it are left unchecked. Whenever a person is said to have relapsed, it means they start experiencing the negative thinking patterns and avoidance they experienced during their darkest times before learning to cope with them.

Dealing with a mental health disorder is quite challenging and complex. Even when you feel like you have complete control over the symptoms associated with it, the risk of sliding back into old habits is quite high.

Signs of Relapse

Here are certain warning signs of a relapse:

• Constant feeling of sadness or anxiety.

• Losing your interest in activities that you once enjoyed. The inability to take pleasure in your hobbies or any other interests you might have, including sex.

• Feeling extremely agitated or restless for no plausible reason.

• Inability to stop thinking about the past or worrying about the future.

• Experiencing an overwhelming sense of guilt or worthlessness.

• Unexplainable bouts of sadness and distress.

• Another common warning sign you can look out for is any sudden change in your eating patterns and appetite. Eating too much, binge eating, emotional eating, or absolute loss of appetite are all warning signs of a relapse.

• An urge to withdraw and avoid all sorts of social situations. Breaking off ties and losing touch with friends or loved ones is also a sign of relapse.

• Any abrupt changes in your sleeping patterns. Sleeping too little, insomnia or an inability to stay asleep throughout the night are warning signs you must not ignore.

• Unexplainable stomach aches, muscle pains, headaches, or any other physical pains and aches can also be warning signs.

• If you notice you have any trouble concentrating or remembering things, it could be a sign of the recurrence of a mental health disorder.

If you notice you are entertaining suicidal thoughts or are engaging in any behavior that harms you, it is a sign you must not ignore at any cost. Whenever you notice any such behavior, it is time you immediately seek help. If you experience hallucinations, dissociation with reality, or anything else along these lines, seek medical help immediately. When left unchecked, these symptoms can quickly spiral out of control and can become a threat to your life.

Once you understand these warning signs, you can be on the lookout for them. Take some time and explain these signs to your loved ones so that they can alert you whenever you seem to be getting off track.

Triggers of Relapse

In this section, let us look at specific triggers you must watch out for to avoid relapse.

Relationship Troubles

Different troubles in a relationship can also be traumatic. They are not physically traumatic but induce mental or emotional trauma. For instance, a parent dealing with empty nest syndrome is quite susceptible to depression. On the other hand, dealing with the death of a loved one can also trigger depression. Any issues in your love life can trigger anxiety disorders. Any form of mental or emotional trauma can lead to a relapse. Any event that induces extreme stress must be dealt with very carefully for the sake of your wellbeing.

Quitting Treatment

Not completing the course of treatment can certainly trigger a relapse in different mental health issues like depression, post-traumatic stress disorder, schizophrenia, or any other illnesses. Usually, people start feeling better, and they immediately stop taking their medicines or stop psychotherapy altogether. By doing this, they don't reach remission and instead enter a phase of relapse. Ensure that you maintain a healthy schedule and lifestyle. Eat healthy and nutritious meals, sleep for at least seven hours daily, exercise regularly, avoid toxic people, and stay away from alcohol or drugs. Start taking care of your physical health, and your emotional and mental health will also improve.

Traumatic Events

Various traumatic events can bring about a relapse, like natural disasters, horrific accidents, or even terrorist attacks. When faced with such trauma, your brain can relapse into old patterns as a coping mechanism. So, whenever you are dealing with a traumatic event, watch out for different signs of relapse.

Addictions

These days, addictions are no longer limited to the consumption of alcohol or drug use. There are different types of addictions, like binge eating; watching excessive television; gaming addictions; and so on. Turning to addictions will certainly provide temporary relief from any symptoms or disorder that you experience. In fact, it is the perfect escape from unpleasant emotions. However, if you start depending on an addiction, it certainly increases the risk of relapse; it could act as a potential trigger. Binge-watching television is believed to be a common trigger for depression, anxiety, and stress.

Certain hormonal changes in women can also trigger depression or other emotional disorders. The brain chemistry regulating emotions is often affected by various hormones. Women nearing the age of puberty, during or after pregnancy, and at the time of perimenopause are more susceptible to developing depression.

If you fall under any of these categories, then it is essential that you seek medical help whenever you notice any of the signs of relapse discussed in this section. Now that you're aware of the different triggers, it becomes easier to avoid them. Whenever you notice a trigger, you can act immediately and take corrective action to prevent a relapse.

Tips to Prevent Relapse

Here are some simple tips you can use to prevent relapses.

Love Yourself

Whenever you start noticing any of the triggers or signs of a relapse, it is time to nurture yourself. Taking care of your overall wellbeing is your responsibility, and you cannot blame anyone else for it. Feeding your senses is essential. When you start noticing any triggers, listen to your favorite music, spend time doing things you enjoy, or perhaps sip on a hot cup of calming tea. Try stimulating the sense of smell and touch. Spend time in nature and allow it to calm you. Engage in some physical exercise to counteract any stress you feel.

Positive Self-Talk

Engage in plenty of positive self-talk. You are your own cheerleader and pep coach. Whenever you feel low or down in the dumps, it is time to give yourself a quick pep talk. Keep reminding yourself that whatever you are experiencing right now is temporary and that it too shall pass, like everything else in life. Just because you feel miserable right now, doesn't mean you will always feel miserable. That is the beauty of life; nothing is constant. So, the next time you start feeling overwhelmed by any negative thought patterns, replace them with positive self-talk. You can tell yourself something like, "I know I will feel better soon," or "I'm just having a bad day, my life isn't bad."

Reach Out

Whenever stress or any other intense emotion gets hold of you, you might experience a desire to isolate yourself. Isolation will only worsen the chances of a relapse. Don't do this to yourself, and instead reach out. If you are struggling with a mental health issue, then let others know about it. Talk to your loved ones about it as openly and freely as you possibly can. Educate them about any issue you are dealing with and provide information about any warning signs they must look out for. It will not only make you feel lighter; they will also get a better understanding of you. There are various support groups you can join where you can meet others who are living through the same things you are experiencing right now.

Preparation

While dealing with a mental health issue, always prepare yourself for a relapse. There will be instances when you might fall back into your old patterns of negative thinking and harmful behaviors. By preparing yourself for this, you can come up with a plan of action that can be used if you relapse. Start by making a list of all the different warning signs that might appear and then draw up individual plans to act on them. If required, don't hesitate to consult a doctor.

Keep Up with the Treatment

For certain mental disorders, the course of treatment includes medication for a specific period. Ensure that you keep up with the treatment prescribed to you and don't cut it short. If you complete the full course of the prescribed medication, it reduces the risk of a relapse quite significantly. After all, the point of treatment is to get better. So, why not complete the course?

Practice Mindfulness

Up until now, you were given different tips about how you can use mindfulness to deal with various mental disorders. It is time that you start practicing it daily and consistently. It certainly takes some time and effort, but it will be worth your while. Start adding different mindfulness practices to your daily life and stick to a routine. You might not notice a change immediately, but you will start to feel better, eventually. Unless you commit yourself to DBT and practice the different techniques prescribed by it, you cannot notice any improvements.

Tweak the 12-Step Approach

The 12-step approach is usually used in programs like Alcoholics Anonymous. As the name suggests, there are 12 steps followed on the road to recovery or sobriety. All the 12 steps might not necessarily work for DBT, but there are certain aspects of this approach that can be tweaked to meet the requirements of DBT. In this section, let us look at certain simple steps you can follow to enhance your ability to regulate any intense emotions and restore your mental health.

• Admit to yourself that you need to learn to regulate your emotions. Understand the importance of emotion regulation and the harm inflicted by emotional dysregulation.

• Consciously search for and fearlessly evaluate yourself. Conduct a moral inventory of yourself without any judgment.

• Admit to yourself and another person the nature of all the wrongs or misdeeds you have committed.

• Make a list of all the people you might have knowingly or unknowingly harmed.

• Start making direct amends to such people whenever you can. However, stop yourself from doing this if you know it might harm the other person.

• Keep taking stock of your personal inventory and admit whenever you go wrong.

- Focus on understanding your emotions, thoughts, and feelings using meditation or prayer.

- Experience an emotional rebirth because of all these steps. Once you complete these steps, it is time you pass on this message to others about dealing with any mental health issues.

When it comes to relapses, there are plenty of steps you can take to prevent them. However, don't get discouraged while dealing with a relapse. It isn't the end of the world. You managed to regulate your emotions using DBT the first time around, and you can do it again. Be patient with yourself and don't rush into it. Don't expect any miraculous results. However, with consistent effort and time, you can see a positive change in your overall mental and emotional health.

Conclusion

Initially, dialectical behavior therapy was developed by Dr. Marsha M. Linehan as a treatment for borderline personality disorder. However, DBT is now being used to treat a variety of mental health conditions and isn't restricted to BPD. DBT can be used to improve your ability to handle any distressing situations in life, without losing control of your emotions, your emotional stability, or resorting to destructive behaviors. It is a great technique for rectifying emotional dysregulation.

The core principles of DBT are based on mindfulness, distress tolerance, emotion regulation, and interpersonal effectiveness. These basic principles come in handy while dealing with difficult emotions. Certain situations in life cannot be changed regardless of how hard you try, which can be a source of immense stress and distress. Learning to cope with such situations and get out of them takes mindfulness. Mindfulness is one of the most important aspects of DBT.

Mindfulness is the ability to live life in the moment, without allowing any thoughts about the past or the future to hijack your thinking patterns. Unless you are mindful of yourself, your emotions, thoughts, feelings, actions and life in general, you cannot lead a happy and stress-free life. It is where DBT steps into the picture. To regain control over your emotions and maintain emotional stability, you must commit yourself. Health isn't just restricted to your physical wellness; it includes your mental and emotional wellbeing too. Unless all these three aspects of your health are in balance, you cannot attain mental peace.

In this guide book, you were provided with the information required to develop and improve important skills that help you focus on your current state, while reducing stress, worries, and PTSD. You were also provided information about effectively counteracting any impulsive behavior using DBT, and tips for dealing with extremely stressful situations in life.

Every technique and tip given in this book is simple to understand and easy to follow. All the advice is curated to help you stay in the present moment, increase your understanding of your emotions, understand your true self, and curb impulsive behaviors. You can do all this even in times of distress. All the techniques covered in this book will certainly help improve your ability to regulate your emotions, while promoting your mental and emotional health.

The key to your emotional and mental wellbeing lies in your hands. The first step toward regaining control of your emotions is DBT. A little consistency and effort are all it takes to master the different techniques of DBT and mindfulness suggested within this book. Once you start following these techniques, you will notice a positive change in your emotional health. So, what are you waiting for? There is no time like the present to get started.

Thank you and all the best!

Resources

https://www.youtube.com/watch?v=wRBw_Iti3Ww
https://www.verywellmind.com/dialectical-behavior-therapy-1067402
https://www.youtube.com/watch?v=ftL7l4KiHag&t=410s
https://bayareadbtcc.com/mindfulness-in-dbt/
https://www.pasadenavilla.com/2019/07/22/mental-v-emotional-health-related/
https://www.webmd.com/mental-health/mental-health-types-illness#1
https://en.wikipedia.org/wiki/Emotional_and_behavioral_disorders
https://www.mayoclinic.org/diseases-conditions/mental-illness/symptoms-causes/syc-20374968
https://www.healthline.com/nutrition/anxiety-disorder-symptoms#section2
https://www.healthline.com/health/depression/recognizing-symptoms
https://www.goodtherapy.org/blog/how-to-create-achievable-goals-for-your-mental-wellness-0822164
https://www.talkspace.com/blog/set-mental-health-goals/
https://anxietyreliefproject.com/managing-anxiety-dialectical-behavior-therapy-dbt/
https://www.anxiety.org/dbt-dialectical-behavior-therapy-compared-to-cbt
https://psychcentral.com/blog/3-dbt-skills-everyone-can-benefit-from/
https://www.behavioralwellnessgroup.com/index.php/articles/125-ten-dbt-techniques-for-anxiety
https://www.clearviewwomenscenter.com/blog/treat-depression-dbt/
https://positivepsychology.com/emotion-regulation-worksheets-strategies-dbt-skills/
https://www.youtube.com/watch?v=lXFYV8L3sHQ

https://www.borderlinepersonalitytreatment.com/dbt-skills-workplace.html

https://www.huffpost.com/entry/managing-work-stress_n_3454501

https://thriveglobal.com/stories/10-ways-to-practice-mindfulness-at-work/

https://www.youtube.com/watch?v=zPopjuKuweg

https://www.verywellmind.com/dialectical-behavior-therapy-dbt-for-bpd-425454

https://www.youtube.com/watch?v=RPgvG13tfAc

https://www.youtube.com/watch?v=aeQwtgFkguU

https://www.getselfhelp.co.uk/distresstolerance.htm

https://www.youtube.com/watch?v=yyH1JLZcVR8

https://www.youtube.com/watch?v=CBopCkdBwsk

https://www.sunrisertc.com/distress-tolerance-skills/

https://www.youtube.com/watch?v=ftL7l4KiHag

https://www.mindful.org/mindfulness-meditation-anxiety/

https://www.intrusivethoughts.org/?topic=mindfulness

https://www.verywellmind.com/relaxation-is-an-essential-ocd-self-help-technique-2510635

https://adaa.org/understanding-anxiety/panic-disorder-agoraphobia/symptoms

https://www.youtube.com/watch?v=_EbqcVH9eVg

https://www.healthline.com/health/how-to-stop-a-panic-attack#close-eyes

https://tinybuddha.com/blog/beat-panic-attacks-3-simple-mindfulness-techniques/

https://www.verywellmind.com/using-mindfulness-for-ptsd-2797588#

https://www.verywellmind.com/dbt-for-ptsd-2797652

https://www.helpguide.org/articles/ptsd-trauma/ptsd-symptoms-self-help-treatment.htm

https://www.verywellmind.com/coping-with-flashbacks-2797574

https://www.medicalnewstoday.com/articles/320269.php

https://www.everydayhealth.com/hs/major-depression-health-wellbeing/factors-can-trigger-depression-relapse/

https://vocal.media/psyche/skills-to-prevent-relapse

https://www.mentalhelp.net/addiction/treatment/mindfulness-based-relapse-prevention-mbrp/

Part 2: Cognitive Behavioral Therapy

A Simple CBT Guide to Overcoming Anxiety, Intrusive Thoughts, Worry and Depression along with Tips for Using Mindfulness to Rewire Your Brain

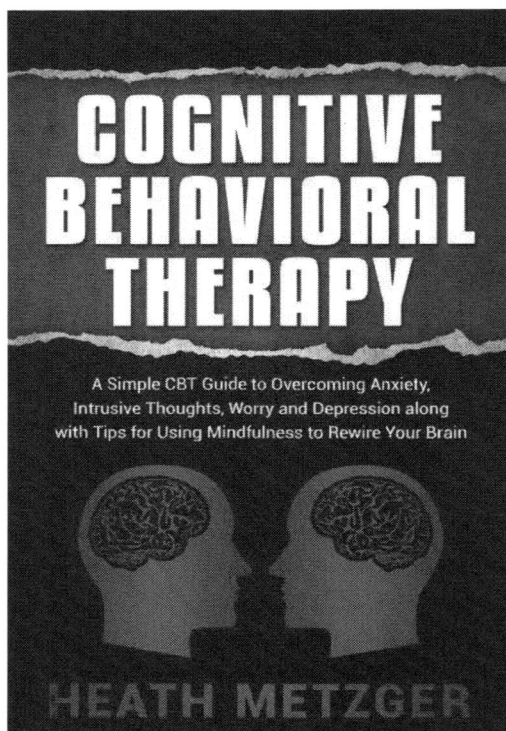

Introduction

Cognitive-behavioral therapy (CBT) has become one of the best methods of psychotherapy for treating several mental health problems. Although it is a relatively new method of treatment, it has gained popularity among many experts around the world. It is important to note that CBT doesn't take away problems; instead, it helps people with mild to severe mental illness to cope and deal with their issues more effectively.

This method of psychotherapy is based on changing and modifying the way you behave. This book aims to create an awareness of CBT, going in-depth to provide you with the mechanisms of action and showing you the inner workings of what CBT is all about.

One of the things discussed is the history of CBT, which will provide you with valid information while you explore the various steps involved in CBT development.

Another essential factor explored is the role of negative thoughts in mental health. What we think about a particular situation affects how we see it, perceive it, how we respond to it, and our behavior towards it. We'll explore the origin of negative thoughts and how a specific pattern of thinking affects our reactions and responses.

Also, we get to see what a typical CBT session looks like and the role of the therapist in helping people with mental health problems. CBT is a therapy that depends on a particular structure; it differs from other treatments in the construction of and the amount of time needed for the procedure. Also, an essential factor differentiating CBT from other methods of psychotherapy is the relationship between the therapist and the patient.

CBT is a goal-oriented therapy, so it is best suited for mental health issues that can be broken down into goals.

Identification of mental health problems is critical, as it helps you to deal with them faster and with better results. Recognition of various health disorders such as depression, anxiety, schizophrenia, panic attacks, excessive worrying, and other mental health illnesses is achieved through observation of the various symptoms. Being able to pinpoint the symptoms of these mental health problems is the first step towards treatment. Many of these mental health problems have similar signs and symptoms, so it is essential to seek the help of an expert to recognize the particular health problem you might be facing.

Goal setting is an essential and fundamental part of CBT. We cannot overemphasize the importance of goal setting because this is one of the first requirements in CBT treatment. Goal setting in CBT has a mechanism that is achieved only by following the correct approach. You will learn how to set realistic and achievable goals in Chapter Three.

Anxiety and worry are prevalent mental health issues around the world. CBT treatment for anxiety and stress consists of asking the right questions and being able to assess yourself correctly. With the help of CBT, you will be able to identify negative thought patterns that often lead to anxiety and worry and to challenge them by replacing them with more realistic ones. In the treatment of anxiety and stress disorders, you and your therapist will work together to come up with various assignments or homework which will help you to apply everything learned in the therapy sessions.

Depression is another major mental health illness that affects many people in this day and age. It is now a common feature of modern-day life and is poised to become one of the most common health problems of our time. There are various types of depression, and identifying the type that you may be experiencing requires professional assistance. CBT treatment for depression often follows the typical goal-oriented, structured plan that is characteristic of

CBT. Unlike other psychotherapy for depression, CBT offers a shorter therapy time, with less chance of relapse. The various tools and techniques derived from CBT sessions can be used long after you've completed the therapy sessions. The treatments remain viable and applicable throughout your life and reduce the risk of relapse.

The relationship between substance abuse and depression is a common topic when it comes to mental illness. Many people who have depression are susceptible to developing addiction problems; this diagnosis is referred to as a dual diagnosis.

Work-related stress is a significant concern, as an increasing number of workers are now experiencing work-related stress at an alarming rate. Because of the growing pressure to achieve more in less time, many people often end up feeling as though they have underachieved, which serves as a significant source of work-related stress. We provide information on the triggers, symptoms, and long-term effect of workplace stress. CBT treatment for workplace stress has been effective in helping many people come up with different strategies for managing it efficiently. CBT helps you to come up with a realistic plan to deal with stressful situations at work by helping you prioritize, keep track of your moods, and feel less overwhelmed at work.

Intrusive thoughts are the leading cause of many mental health problems such as anxiety, depression, Obsessive-Compulsive Disorder (OCD), and others. CBT treatment for intrusive thoughts will help you to identify these thoughts and the triggers responsible for them.

The relationship between CBT and mindfulness is discussed in great detail. The histories and similarities between mindfulness and CBT will help you to understand their usefulness in tackling many mental health issues.

The combination of CBT and mindfulness has produced results that make MBCT an invaluable tool in dealing with mental health problems such as depression and anxiety. MBCT applies the principles of both CBT and mindfulness. If you are tired of various other methods of treatments that just haven't worked for you and you're looking for something more refreshing with a higher rate of success, then CBT might be for you.

Chapter One: Why Use CBT?

Cognitive-behavioral therapy (CBT) aims to improve your general mental and emotional state with the use of a practical approach. It helps you understand complex and challenging cognitive distortions in your thoughts, feelings, and attitudes to improve the ability to function and the overall quality of life. It was designed initially to treat depression and other mental illnesses, but it is now being applied extensively in the treatment of various psychogenic problems ranging from sleep disorders to drug and alcohol abuse, depression, severe mental illness, eating disorders, and anxiety. By paying attention to beliefs, attitude, and lifestyles, it aids you in identifying thinking patterns responsible for ineffective behavior and negative moods. Identifying these patterns improves your emotional control and helps you to develop coping strategies against emotional difficulties that lead to depression.

CBT also features an essential advantage over several other treatment methods; it is a short-term treatment compared to others. For most emotional problems it has a treatment plan ranging between five to ten months. It features a smaller amount of sessions per week, each lasting for approximately 60 minutes. During the sessions, you work with the therapist to identify the psychological triggers that impede typical day to day function. Throughout the therapy sessions the therapist introduces several techniques that can be applied when the need arises, and these principles can be used over a lifetime.

CBT is based on the principle of the addition of behavioral therapy to psychotherapy. Behavioral therapy focuses on the role our problems play in affecting our thoughts, behavior, and general lifestyle. Psychotherapy pays close attention to the beginning of our thinking pattern in childhood and how the importance we place on things affects our response to them. It then aids people in challenging those automatic thoughts, reactions, or beliefs that come when triggered by specific situations, using appropriate strategies to modify their behavior to more positive responses and thought.

The History of Cognitive Behavioral Therapy

Aaron Beck is credited for inventing Cognitive Behavioral Therapy in 1960. He was a psychoanalyst, and while in analytical sessions he saw that his patients had conversations with themselves, similar to those between two people. But his patients never gave a full report of their thoughts.

For instance, during a session of therapy, you might be thinking to yourself, "He (the therapist) doesn't understand anything I'm saying," and you might end up feeling annoyed or displeased as a result of this thought. Consequently, you could respond to the previous thought with another one: "Maybe it's my fault for not being forthcoming with what I feel." The next thought might end up changing how you felt previously due to the first thought.

Aaron finally came to recognize the importance of the relationship between feelings and thoughts. He then coined the term "*Automatic Thought*" to explain emotion-filled ideas that come up from time to time in our minds. He concluded that people are not always aware of these thoughts but can still be taught to identify and report them. He concluded that the key to someone dealing with such difficulties is recognizing and understanding the roles these thoughts play on their emotions and behavior.

Because of the importance of the relationship between thoughts and behavior, Aaron came up with the term "*Cognitive Therapy*" for this method of psychology. Today, it is recognized as Cognitive-behavioral therapy (CBT). The balance between psychotherapy and behavioral treatment can be adjusted to suit individual needs; this has led to the founding of many categories of CBT. CBT has undergone several scientific trials by various teams, and its range of applications has increased through the years.

The Role of Negative Thoughts

The theory supporting CBT is that events alone are not what is responsible for what we feel or how we behave, but also the meaning and importance we place on them. For example, if one experiences a barrage of negative thoughts it could distort their perception and lead them into believing what is not so. They may cling to this distorted set of thoughts and beliefs and fail to accept anything to the contrary.

For instance, due to depression, a student might think, "I'm incapable of getting through school today because nothing will go right; everyone hates me, I have no friends, and I will be all alone." As a result of believing this thought, they might claim to be sick to avoid going to school. By responding and letting this thought affect his or her behavior, he or she takes away the chance of discovering if her prediction is going to be wrong. If the student had ignored this thought and took the opportunity of going to school, something might have been different, or things might have gone better than their prediction. But they choose to stay at home, dealing with even more negative thoughts like, "I've missed a lot today, and I'm so lonely." Views like this could lead to the student feeling even worse and reduce the chances of them going to school the next day. Situations like these are the beginning of a downward spiral. Vicious circles like these often apply to many other problems.

Where Do These Negative Thoughts Come From?

According to Aaron, every individual establishes their thinking pattern during childhood, and they go on to create a reflexive or automatic way of thinking that remains fixed. For instance, consider a situation where the parents generally neglected their child (except in times of need), but the child does well in school. In this instance, the child will end up thinking, "I have to be at the top of my game in school, so I do well, or else my parents will reject me." The child ends up coming up with a rule for his/her existence (referred to as dysfunctional assumption). This rule might be distorted, but it might also enable the child to work harder to do well in school so as to gain more attention from their parents.

This pattern leads to several trains of thought known as dysfunctional thought patterns, which are activated or triggered when something beyond the control of the child happens. When this occurs, the established automatic model of thoughts takes center stage in the child's mind, precipitating thoughts such as, "I am a failure and have no reason for existence."

The role of cognitive-behavioral therapy is to help people in similar situations to understand what is happening. It helps them think outside their established reflexive lines of thought. So, in the case of the student who is worried about how alone they would feel in class, CBT encourages and aids them to examine real-life situations and see what happens. When the student takes the chance and puts themselves in a more realistic real-life state, they might experience things that would go better than they had expected; they may meet someone who shares the same life-view and make a friend for a change, so that they can feel less lonely.

It is an absolute fact that things don't always go as planned. Still, when your mind is unstable, your thoughts, predictions, and interpretations will be distorted; you don't see things clearly, and you will have greater difficulty dealing with worse situations. CBT helps you correct distorted views and interpretations you might have of various circumstances.

What is CBT Treatment Like? What to Expect in CBT Sessions

CBT is different from other types of therapy in some essential ways. If you are considering CBT as a course of treatment, it might help to have an idea of what to expect in CBT sessions.

CBT Sessions

In the first session, the primary purpose is to come up with an assessment of the situation. The meeting allows you to explain in detail the problems you are experiencing. In this session, the therapist will try to get a full picture of the essential factors needed in the care. The therapist then determines if they are the right fit for you in helping you deal with your problems, or if they need to refer you to someone else. At the end of the first session, the therapist is usually able to come up with a treatment plan listing various interventions needed for the therapy. In some cases, it takes more than one session to create a comprehensive treatment plan. After crafting the schedule, the therapist helps you understand if the treatment plan is perfect for you.

After the assessment, treatment begins in subsequent sessions to deal with the objective of the treatment plan. In each session, time is given to solving problems, in contrast to the traditional treatment that involves a lot of talking about the problems.

At every session, CBT maximizes the use of time for greater effectiveness. Every session starts with a little check-in to determine how effective the treatment plan is in solving the underlying problem. A quick recap of the last session and homework follows this check-in. After going through this, the rest of the meeting is spent dealing with the agenda of the day.

Concluding Sessions

When you've met the goal(s) of the treatment plan, the therapist reduces the frequency of the sessions. Other conventional therapy sometimes lasts for three years. However, CBT lasts for just a few months. The reason is that the CBT design aims to make you your own therapist by providing you with the tools to cope with any situation. Therapists schedule the last phase of CBT sessions less frequently, giving you more time to apply the skills and tools you have learned to real-life situations and to build up your confidence in your ability to deal with subsequent problems that may arise.

At the end of each session, the therapist often gives homework; this is a vital process as it helps people master the tools or skills acquired during the sessions. Homework varies depending on the nature of the agenda of a particular therapy session. For instance, you might be asked to keep records of incidents that trigger emotions of depression or anxiety. This assignment might be imperative to examining the thoughts generated due to the events. The next task could be to apply specific skills learned during the therapy sessions to deal with similar situations.

How CBT Differs from Other Therapies

The significant difference between CBT and other therapies lies in the relationship between you and the therapist. An innate drawback of most treatments is that they might create a sort of dependency on the therapist. As a result, you might come to view the therapist as all-powerful and all-knowing. With CBT, this is not the case.

In CBT, the relationship between you and the therapist is more or less an equal one, more like a business arrangement, more practical and problem-focused. Your therapist always asks questions on your views about the therapy. Aaron Beck described this relationship with the term "*Collaborative Empiricism*". It stresses the importance of you working together with the therapist in coming up with ideas to apply to your issues.

Benefits of Cognitive-Behavioral Therapy

The National Association of Cognitive-Behavioral Therapists (NACBT) have described the foundation of CBT as being built on the theory that our behavior and feelings occur in response to our thoughts, and they are not caused by things like situations, people, and events. With this fact, the advantage is that we can alter our views to change our feelings and act contrary to a case or circumstance.

Benefits of CBT include

Ability to identify negative emotions and thoughts

In cases of addiction, deterring relapse

Help in anger management

Coping with grief and loss

Managing chronic pain

Overcoming trauma and dealing with PTSD

Overcoming sleep disorders

Resolving relationship difficulties

Who Can Benefit from CBT Treatment?

CBT is often a suitable treatment for people with specific problems; for people who are just experiencing feelings of unhappiness and unfulfillment but lack the symptoms that prevent them from getting through everyday life, CBT may be less helpful.

CBT is useful for the following problems

Anxiety and panic attacks

Intrusive thoughts

Depression

Worrying

Anger management

Child and adolescent problems

Chronic fatigue syndrome

Chronic pain

Mood swings

Habits, such as facial tics

Eating Disorders

Obsessive-compulsive disorder (OCD)

General health problems

Drug or alcohol addiction

Phobias

Sleep disorders

Sexual and relationship issues

Post-traumatic stress disorder (PTSD)

Mindfulness-Based Cognitive Therapy (MBCT)

Founded by Zindel Segal, Mark Williams, and John Teasdale, MBCT was designed initially to deal with depression, but it is now applicable to a wide range of problems.

MBCT is a psychological treatment that combines techniques of cognitive-behavioral therapy (CBT) with strategies of mindfulness to help individuals better understand and manage their emotions and thoughts. MBCT generally helps individuals achieve alleviation from feelings of distress.

Chapter Two: Identifying Mental Health Disorders

The World Health Organization has given us a clear definition of good mental health; it says that it is a state of well-being where an individual realizes their abilities, can work productively, can deal with the regular, day-to-day stress of life, and can contribute to his or her community.

Mental health disorders include a wide range of various problems, with different signs and symptoms but are characterized by a combination of abnormal emotions, behavior, thoughts, and relationships. Instances of mental health disorders are depression, schizophrenia, intellectual disabilities, and complications due to drug abuse and anxiety disorders. Most of these disorders can be successfully treated.

Mental health disorders can affect relationships, daily life, and sometimes physical health. According to experts, everyone has the potential to develop mental health disorders, regardless of age, ethnic group, gender, or financial status.

Cognitive-behavioral therapy (CBT) used either alone or together with other treatments has been applied to treating several mental health disorders. Besides treating mental health problems, CBT can be used to deal with stress.

We are going to focus on identifying some of the most common mental health disorders, such as anxiety disorders, worry and depression, mood disorders, and schizophrenia.

Anxiety Disorders

Anxiety disorder occurs as a result of extreme fear or anxiety, which can be triggered by specific situations, sounds, and objects. Individuals with this disorder try to keep away from the triggers of their distress.

Some anxiety disorders include:

Panic Disorder

This disorder occurs when you experience sudden paralyzing terror or have a sense of impending disaster.

Phobias

These range from simple phobias such as social phobias (fear of being judged by others) to agoraphobia (fear of being unable to get out of certain situations) to disproportionate dread of objects.

Obsessive-Compulsive Disorder (OCD)

OCD is an anxiety disorder characterized by a feeling of intense urgency, compulsions, and obsessions. For instance, having an intense urge to wash your hand repetitively.

Post-Traumatic Stress Disorder (PTSD)

PTSD occurs in the aftermath of a traumatic event when someone experience or witness something frightening and horrible. During this event, you may have felt threatened or that you had no control over what was happening.

Some signs and symptoms of anxiety include:

Excessive Worrying

One of the most common symptoms of an anxiety disorder is worrying. This emotion is normal in people, but here, the amount of fear in a situation is distorted or not proportionate to the trigger. This symptom is a sign of anxiety disorder if it continues for six months or more and the worry becomes difficult to control, interferes with your daily tasks, and makes it hard to concentrate.

Feeling Agitated

The feeling of anxiousness causes overstimulation of the nervous system which results in a series of connected events that occur throughout the body, such as sweaty palms, a racing pulse, dry mouth, and shaky hands. All these symptoms occur when the body believes it is in immediate danger, so the nervous system prepares the body for it. Blood is diverted away from the gastrointestinal tract to your muscles in preparation for the reaction of fight or flight. Your heart rate increases, and all your senses are alert. These changes are necessary to keep you safe in times of actual danger, but in the case of anxiety disorders, the threat is all in your head. People with anxiety disorder often have a hard time calming their feelings of agitation.

Panic Attacks

A primary symptom of anxiety disorder is a panic attack. This symptom manifests as an overwhelming and intense feeling of fright.

Following the panic attack is an increase in the heart rate, shortness of breath, shaking, tightness of the chest, nausea, sweating, and loss of control. If panic attacks occur frequently, then it may be a sign of anxiety disorder.

Trouble Falling or Staying Asleep

Sleep disorders usually accompany anxiety disorders. You might have a hard time falling asleep and/or staying asleep. According to some studies, children who have insomnia are more at risk of developing an anxiety disorder later in life. This symptom is strongly related to anxiety, and treatment of anxiety disorders improves sleep as well.

Restlessness

Restlessness is a common sign of anxiety disorder, especially in teens and children. People with this symptom tend to be on edge and have an uncontrollable urge to move. This symptom might not occur in everyone with anxiety, but it is an essential symptom of anxiety disorders. Experiencing restlessness for more than six months is a sign of an anxiety disorder.

Irrational Fears

Intense fear of certain things such as enclosed spaces, heights, and spiders could be a sign of phobia. Phobia is the irrational fear of a particular situation or object. This fear impairs your ability to carry out everyday functions.

Common Phobias Include

Phobia of animals

Situational phobias

Phobia of the natural environment

Phobia of blood/injections/injury

Avoiding social situations

You might be displaying indications of social anxiety disorder if you find yourself having a fear of social events, fear of being judged or examined by others, or the fear of public humiliation.

Avoiding Social Situations

Social anxiety is common, even among adults, at some point in their lives. It is commonly developed during childhood and is usually common in early teens. An individual with this symptom tends to appear extremely quiet and shy when meeting new people or in group settings; though they may hide this on the outside, on the inside they have extreme fear and distress. Some people can appear standoffish or snobby as be a result of high self-criticism, depression, or low self-esteem.

Difficulty Concentrating

People with anxiety disorders often have a hard time concentrating. According to some studies anxiety impairs working memory, making it impossible to hold on to short term memories. However, it's important to note that not being able to concentrate is a sign of other disorders as well, such as depression, so it not an exclusive symptom of anxiety disorder.

Depression

Feelings of unhappiness are different from that of depression. Depression is a mental health disorder that is more complicated. It is a significant mental health disorder that leads to difficulty in carrying out simple daily tasks.

Feelings of sadness are part of our everyday life. Still, when you start to experience emotions like despair and hopelessness that won't go away, you risk a downward spiral into depression.

It is essential to take note of the signs and symptoms that point towards depression to determine if they are the natural feelings of sadness that most people tend to sometimes go through in life.

Hopeless Outlook

A primary symptom of depression is the way people with this disorder perceive life. They have a pessimistic or helpless view of life which prevents them from functioning. They may also have feelings of worthlessness, self-hate, and intense guilt, believing everything is their fault.

Loss of Interest

People with depression often cannot feel pleasure or enjoy things. They experience a loss of interest in everything in general, even those things they once loved. Loss of sex drive and/or impotence are primarily associated with depression.

Increased Fatigue and Sleep Problems

As a result of the inability to enjoy the things you love, you feel a never-ending sense of fatigue. You might feel a lack of energy or have an excessive urge to sleep. Depression can also be associated with insomnia.

Anxiety

Depression and anxiety can occur simultaneously, and often go hand in hand. It is essential to note that not all forms of anxiety are symptoms of depression. Some things that can be associated with anxiety include:

Feelings of dread

Fast and increased heart rate

Nervous breathing

Restlessness and tension

Overthinking and lack of focus

Increased sweating

These symptoms can be managed with CBT.

Irritability in Men

Depression tends to be different for men and women. Depressed men usually have characteristic behaviors such as irritation, substance abuse, and misplaced anger. Unlike women, men with depression find it difficult to recognize depression and seek treatment.

Changes in Appetite and Weight

Weight fluctuation is one of the primary characteristics of depression, and it is different from one individual to another. In some, it leads to a gain in weight, while in others, it can lead to weight loss. Depression is known to affect appetite, and it can differ from one person to the next as to whether it pushes them to overeat or not eat at all.

Uncontrollable Emotions

When depressed, you might get periodic outbursts of anger. Your feelings can be all over the place; you can go from feeling an intense amount of anger from the smallest trigger at one moment, and the next you may find yourself wallowing in self-pity or crying for no apparent reason.

Looking at Death

The leading cause of suicide is depression. A large number of people die each year as a result of depression that has gone untreated. They often see it as a way out of their problems because, from their point of view, there's no other way out. They may talk about how they often consider taking their own lives. If you think someone is at risk of committing suicide as a result of depression, it is crucial to get them help as soon as possible.

Mood Disorders

This mental health disorder is primarily concerned with the emotional state of a person. The person experiences extremes of highs and lows, of happiness and sadness, or sometimes even both.

Humans have the natural ability to alter their moods on the basis of their surrounding environment. However, in the case of mood disorder, the ability to handle routine or normal daily activities is interrupted and challenging.

Common symptoms of mood disorders include:

- Feeling sad almost all the time or nearly every day
- Feeling worthless or hopeless
- Lack of energy or feeling sluggish
- Gaining weight or losing weight
- Extremes of appetite (high or low)
- Lack of sleep or oversleeping
- No interest in activities that you normally enjoy
- Frequent thoughts about death or suicide
- Difficulty concentrating or focusing
- High-level energy
- Fast speech or movement
- Perturbation, unease, or touchiness
- Risk-taking behavior such as driving recklessly or drinking excessively
- An abnormal increase in activity or an atypical urge do too many things at once
- Contending thoughts
- Feeling apprehensive or edgy for no apparent reason

Schizophrenia

Schizophrenia is a serious and debilitating mental disorder wherein people have a distorted view of reality. It is usually characterized by hallucinations, resulting in the inability to carry out daily functions. Schizophrenia requires lifelong treatments to manage. In the early stages, medical care is given to manage the symptoms and improve lifestyle.

Features of this mental health disorder involve a range of emotional and behavioral problems, and the inability to distinguish the difference between what is real and what isn't. Common signs and symptoms vary from person to person. Symptoms like speech impairment, hallucinations, and delusions cause great difficulty in functioning effectively.

Symptoms of schizophrenia include:

Delusions

People with schizophrenia have difficulty in perceiving reality, as some of their beliefs are false and are not based upon reality as most people see it. For instance, people with schizophrenia may believe that they are being harmed and harassed, that there is a significant disaster coming, that someone is in love with them, that they have attained exceptional fame, or that they have gotten a dream job. Most people with schizophrenia have these kinds of delusions.

Hallucinations

Hallucinations are another significant sign of schizophrenia. It involves hearing and seeing things that are not there. This symptom could occur in any of the senses, but the most common is the patient hearing voices in their head. The next is seeing things that aren't there.

Disorganized Thoughts (speech)

Schizophrenia is also characterized by uncoordinated expression as a result of disordered thinking. This makes it challenging to communicate and pass information. "Word salad" is the term used to describe the situation where they utter meaningless words as they attempt to express themselves.

Extremely Disorganized or Abnormal Motor Behavior

People with schizophrenia also may exhibit childlike silliness and compulsive agitation. They find it hard to accomplish tasks as they lose track of the goal or purpose of a given job. Other common behaviors include resistance to instructions, strange postures, excessive and unproductive movements, and refusal to respond while being talked to.

Negative Symptoms

This symptom is characterized by the inability to function properly. For instance, appearing emotionless (failing to make eye contact, having no facial expression, and talking in a monotone). People with this condition experience a loss of interest in various activities, social withdrawal, and an inability to feel pleasure.

Over time the person experiences periods of heightened symptoms and periods of remission; however, some symptoms may persist regularly.

For men, studies show that schizophrenia usually begins between their early to mid-twenties, while for women, it usually starts in the late twenties. Schizophrenia is rare in children and for individuals above the age of 45.

Symptoms in Teenagers

Though the symptoms of schizophrenia are the same in adults and teenagers, the symptoms are more difficult to recognize in children. This is because early symptoms are similar to the expected behavior of teens in general, so schizophrenia in teens can go unnoticed for some time. Some of these symptoms include:

- Keeping a distance from friends and family
- Reduced performance in school
- Sleeping disturbance
- Bad mood
- A feeling of doing nothing

Compared to schizophrenic adults, teens are less likely to have delusions and more likely to have visual hallucinations.

Severe symptoms like extreme shock, loss of control, loss of touch with reality, hallucinations, urge to self-harm, and suicidal thoughts in mental health disorders are considered psychological red flags. If by chance, you notice any of these symptoms, you must seek professional advice immediately.

Chapter Three: Goal Setting: Your Starting Point to Mental and Emotional Wellness

The CBT method has its foundation based on the collaboration between you and the therapist to design strategies and structure to maintain focus. In other words, CBT is designed to be goal oriented. The whole point is to make the reason for the therapy relevant to you. The goal is emphasized by the therapist to give you a clear view of what you want instead of what you think you want.

The Therapeutic Relationship in CBT

In CBT, the collaborative relationship between you and the therapist is important; it is referred to as a "therapeutic relationship". In this relationship, the therapist acts more like a mentor or guide as opposed to the instructor in other types of therapies where you are told what to do. In CBT, the therapist acts in a supporting role, pushing and encouraging you towards exploring new options on how to control your thought process to manage your behavior and feelings. This relationship plays a vital role as you work with the therapist to come up with realistic goals to determine what you hope to accomplish.

How Goals Work

Creating goals helps you to focus on what is relevant and essential. It also helps you develop a vision of the place you want to be in your life, or how you would ideally want your life to be. Setting a goal increases your effort or decreases your energy on specific activities to help you come up with strategies to achieve the goal.

Generally, goal setting in mental health is an essential step used to recover from any common mental health problems such as anxiety and depression. Cognitive Behavior therapy is the first step in overcoming such issues.

Approach to Goals

There are many different ways or paths to goal setting. In CBT, the "S.M.A.R.T." way is one of the most used strategies. This method gives you a vivid and clear picture of your goal or what you hope to accomplish. It aids you in maintaining your zeal or motivation for achieving your purpose. Below is the full meaning of the acronym "S.M.A.R.T."

S-Specific

Being specific means that your goal is clear and focused on what you want. This ensures that you avoid generalization and increases the chances of accomplishing your desired goal.

M-Measurable

Being a measurable goal enables you to quantitatively and qualitatively create margins for what you want. It gives you a concrete criterion toward achieving each goal set. You need to ask yourself, "How many?" and "How much?", and, "How will I know when my goal has been attained?"

A-Achievable

Your goals need to be achievable and feasible. Questions like, "How are you going to go about achieving your goals?" and "What can you do to make meeting your goals possible?" need to be asked.

R-Realistic

This is one of the most important aspects of goal setting. Your goals need to be within the limit of what you think you can do within the time frame you have given yourself. Is the goal achievable given your current circumstances? Though setting big goals can serve to motivate you towards working hard, it can be disappointing when they are not achieved because they are set too far beyond your capabilities. This can leave you feeling even worse.

T-Timely

The time frame for achieving each goal must be within a realistic limit. This helps you to avoid procrastination that could lead to you giving up on your goals.

Below are two examples of achieving your goals with this approach in CBT.

A person not currently exercising has the desire to exercise frequently. What they might be able to put together using the SMART approach is:

S-Specific: Every 20 minutes, I will jog around the nearby park.

M-Measurable: I will keep a diary or record the number of times I go jogging and for how long.

A-Achievable: I will ask a friend to join me on each jog so that I am obligated to go out.

R-Realistic: 20 minutes is more than enough time to have a good morning jog around the park, so it is enough to get my juices flowing, and with my friend with me the run will be super fun.

T-Timely: I will keep up this routine for a month, and after that I will have a review of how much success I have had in achieving this goal.

The second example is Michael, who experiences panic attacks as a result of severe anxiety. Michael's academics are greatly affected as a result of the stress, even though he is a good student. He finally gets through college and begins work, and now deals with anxiety and occasional panic attacks when interacting with his colleagues. He decides to set the goal of coming up with a way of dealing with the anxiety disorder to improve his mental health. Applying the smart method, this is what he can come up with:

S-Specific: he has concluded that he needs to reduce his anxious feelings and to decrease his panic attacks.

M-Measurable: to make his goal more measurable, he decides to keep a daily record of his moods and grade the level of his anxiety on a scale of 1-10. He will do this each time he experiences the slightest hint of a panic attack. This is so he can get the data he needs to determine if there is any change with time.

A-Achievable: the goal of becoming less anxious and more confident is one that many people have achieved, so it is a pretty easy goal to achieve.

Realistic: the goal is a realistic one that is within the limits of his power. After seeking help and gathering the necessary information needed via the Internet and other means of research, he confirmed that his goal is achievable. He found a therapist and was advised that though his goal is possible, it requires a lot of personal hard work, but he can reduce the rate and level of anxiety he feels to manageable levels.

T-Time

With the daily record-keeping he hopes to see noticeable changes by the time he analyses his data after a year and a half. At that time, he hopes to feel more confident in dealing with more challenges at the workplace.

In these examples, both people were able to come up with a well-thought-out goal backed up with a plan that was both realistic and within achievable limits. In time, both goals have a high chance of becoming reality. The S.M.A.R.T. approach provides you with a step-by-step plan that provides the most comfortable possible route toward achieving your goal.

Role of Homework in Achieving Your Goals

As mentioned in Chapter One, homework is an essential aspect of CBT. It is following the strategy and plan set out to make your goal a reality. If you want to get better and improve your mental health, this part of CBT is a must.

The activities or assignments will depend mainly on the therapeutic relationship between you and the therapist. After every session, you and the therapist come up with jobs that will provide you with opportunities to put what you have gained so far in the sessions into practice. The therapy itself is an excellent setting to gain useful insight that you may not have been able to see without guidance. So basically, the purpose of these assignments is for you to align and put the things you've learned into practice. This helps you gain real-life experiences and come to an understanding of just how much better you can become at controlling your thoughts and behaviors.

In addition to doing these assignments between sessions, you will also have to keep a record of your findings on the CBT worksheet.

An example of an assignment is keeping records of your thoughts in response to certain different situations to help you identify those objects, things, locations that serve as triggers for specific unhealthy thought processes.

Steps in Achieving your Goals

Breaking Down Goals

If your goal consists of many working parts, then breaking it into smaller segments is not a bad idea. It allows you to have a sense of accomplishment that maintains your focus, and therefore, you don't have to feel overwhelmed.

Sometimes the goals of people with a mental health problem are ambiguous and unspecific, such as:

I want to feel happier

I want to feel better about myself and gain self-confidence

I need to stop being anxious all the time and become more relaxed

All these goals are ambitious, and there is nothing wrong with them. However, they are not specific enough. They are too broad because you don't know where to start; what do you do, how do you accomplish it, and how do you know how far you have gone in achieving the goal? These are questions you need to ask yourself.

To help you to establish a reasonable and realistic goal, below are a few steps to help you to break down your goal into smaller achievable goals.

The first step is asking yourself - how do you want to live your life once accomplishing the goal?

With this question, you move a step closer to achieving your desired state of mind, even though it doesn't do much to rid you of anxiety, low self-esteem, or depression. Some questions you can ask yourself at this stage are:

· What do you want to experience to feel better about yourself?

· What statements would you begin to make when you have succeeded in increasing your self-confidence?

· What engagements will you have with your increased self-confidence? What will be your approach to life?

The next step is for you to find real answers to your questions.

You will have to recognize specific thoughts and behaviors as guides. Below are some probable answers:

-With me being happier, I can spend more time with my friends and colleagues. I will have at least three social outings with them as a sign of me feeling more satisfied.

-When talking to myself, I will congratulate myself on handling certain situations. I will feel proud of myself for having those social outings and for being happier.

Set a Time Frame for Achieving Your Goals

You will need to keep track and record your progress on a daily or weekly basis. You will have to keep track of how many times time you go out with your friends and how many times you congratulate yourself on a job well done, as well as record things that prevented you from achieving your goals.

The routine check enables you to see your progress and gauge how far you have come or how close you are to achieving your goal. You will also need to find a way of congratulating yourself on celebrating your accomplishments after each routine check. This will serve to increase confidence in yourself.

Take a break to look at how much progress you have made.

Take a short break to look back on how much progress you have made so you can see the bigger picture. As you examine the data, pay attention to how consistently you've been able to fulfill your goals. Do you feel any different from how you felt before? These questions help you to determine what you have been doing right, and if there is anything you should be doing differently. This reflection creates room for you to analyze your newly formed routine of behavioral patterns so you can see if it is going to affect your lifestyle while allowing you to appreciate how far you have come in achieving your goals.

You might experience some obstacles that get in the way of you achieving your goal. One of the significant factors of CBT assignments and keeping a record of your progress is to allow you to recognize the obstacles that hinder you. This allows you to come up with effective strategies to tackle them. For instance, you might discover new ways of calming yourself during a panic attack by coming up with helpful thoughts that serve to soothe and stabilize you. Identifying these obstacles also helps you to come up with methods of ignoring and abandoning specific thought patterns that keep you from achieving your goals.

Goal setting and using a workbook

A goal-setting workbook is the perfect tool to help someone achieve his or her dreams and goals. It allows you to keep track of your progress and get things down on paper. A workbook allows you to:

· Keep a record of all your accomplishments

· Keep track of things you want to avoid doing

· Identify barriers keeping you from achieving your goal

· Keep a record of things that help get you through certain situations

· Set up long-term and short-term goals.

- Recognize areas needing improvements
- Come up with healthier routines

Chapter Four: Anxiety and Worry: CBT Techniques to Reduce Both NOW

Stress and worry are on the rise in our everyday life. Being always connected to the news, social media, and other aspects of the Internet has served to increase the risks of anxiety and worry. The increased pressure to achieve more, and uncertainty about our finances and career have also helped to contribute to increasing fear in people's lives. People now train themselves in the act of multitasking to do more work to meet daily needs, and this has only fostered anxiety. The mental health of the general population is at risk; now more than ever, the need to find a solution to the problems created by our society is on the increase.

Excessive worry, more often than not, leads to anxiety. Some people argue that worrying helps them to prepare for the unexpected, and while this may be true for certain people, for others it only leads to more problems.

Anxiety, worry, obsessive thoughts, and panic attacks are treatable. These mental health problems and other similar ones can all be managed to a level that allows you to lead a healthy life. Although medications for anxiety are quite useful to manage stress and worry, they only go as far as treating the symptoms. Anxiety therapy is the most effective course of treatment for anxiety, excessive fear, and panic attacks because, unlike medication, therapy goes as far as addressing the underlying cause of the mental health problem. It helps you discover the roots of your worries and fear while helping you come up with strategies on how to overcome them. It shows you how to look at the causes of your worries and anxiety in less frightening ways.

There different types of anxiety disorders, and each one is considerably different from another. This means that the particular therapy assigned for each one must be designed for the specific diagnosis and symptoms. The treatment plan for anxiety attacks will differ from those for a person with obsessive-compulsive disorder (OCD). The duration of time needed for therapy will also be dependent on the type and severity of anxiety disorder.

Many therapy techniques and types have been applied in treating anxiety disorder and excessive worrying, but the most effective approach is Cognitive Behavioral Therapy (CBT). Sometimes CBT is used together with other necessary therapy techniques depending on the need of the individual.

You must determine if your worry is either just the usual worry everyone has or excessive to the point of preventing you from leading a healthy life. Self-evaluation is essential to determine if your anxiety is a mental health problem. You can start by asking yourself the following questions.

Possibilities of Your Worries Coming True

What are the possibilities of what you are worried about happening?

You need to recognize what you are anxious and afraid of. For instance, if you are meant to give a speech, you might be concerned about people making fun and laughing at you. If you are worried about meeting with someone, it may be that you are afraid of what the person might say or that you may be rejected. If your worry is related to making mistakes at work, it might be that you are worried about being fired.

You have to be able to figure out what frightens you if you want to have any hope of overcoming anxiety. Most times, recognizing what you are terrified of will make you realize how baseless your fear is.

After determining the source of your fear, you need to determine the probability of your concern coming to pass. You need to logically analyze the times you have been in a similar situation and determine the number of times the worst-case scenario your fear predicts has come to pass. For instance, what is the likelihood of people making fun of you while you are giving a speech? Are there things you could do to decrease the probability of this happening? All these questions can make you realize that there are actual ways to influence the situation favorably.

Is the Best-Case Scenario Possible?

Sometimes people assume the worst-case scenarios when it comes to almost every situation. They have become experts in imagining the worst in everything, and as a result, they have forgotten that something contrary to their imagined scenario is possible. It is okay to consider the worst-case scenarios, and it might even prove to be useful, but when we forget about the best-case scenarios in every situation, then we might be taking our worries too far. You have to determine the two possible extremes of a scenario and then consider the most likely one. Most times, your mind wanders toward the more extreme scenarios while, in most cases, that doesn't happen. If you have difficulty determining the most likely scenario, it might do you good to imagine a situation with a mixture of both good and evil.

How Many Times Has Your Prediction Happened?

This is another useful way to determine if your actual worry is worthwhile by having a count of how many times the predicted worst-case scenario became a reality. If you are still okay, despite getting into a similar situation so many times over the years, this might mean your worry is baseless. Even if the worst-case scenario has happened before, you can compare the number of times it has happened to the number of times it hasn't. This can make you realize your fear might be unnecessary.

What Can You Do to Cope, even if the Worst Comes True?

Most times, you will see that your worry seems to go only towards the worst-case outcome. What comes after that seems not to be what worries us. It may be helpful to add "What happens next?" to your list of worries. This might help you figure out what you will do to cope if the worst-case scenario does happen. If you do get laughed at or made fun of when giving your speech, will you go home and sit in front of the TV all day or maybe sleep all day? Whatever the case might be, if you can extend your worry toward imagining a coping strategy, you will realize that you will be fine even after the worst-case scenario.

What Good Does Worrying About Every Situation Do?

This is the final question that you need to examine. Does worrying about the situation change the outcome? Does it make you any better at handling the outcome, or does it just make things worse? Sometimes worrying might serve to get you pumped up for the situation and prepare better. But too much worrying might do more harm than good. For instance, it might interfere with your preparations for an event; you might end up believing the worst-case scenario instead of looking at facts. No matter how much worry you put into the specific situation, the outcome cannot be influenced by it. So instead of worrying, you might as well spend your time doing things to prepare yourself better. Your worry might be self-defeating but focusing your mind and training your brain to address specific concern, you will become less likely to be taken over by your anxiety and worries.

This method of self-examination to determine if your worry and anxiety are doing you more harm than good is derived from Cognitive Behavioral Therapy. CBT has been proven to be the most effective method of therapy for obsessive worry and anxiety disorders.

Treating Obsessive Worry and Anxiety With CBT

CBT is based on the fact that our behaviors and feelings are dependent on our thoughts. People with anxiety disorders have a negative way of thinking that serves as a source of negative emotions, fear, and anxiety. CBT treatment of worry and anxiety helps you to recognize those negative thought patterns and correct them.

CBT for Anxiety - Thoughts to Challenge

This is referred to as cognitive restructuring. In this process, the negative thought pattern and beliefs of the individual with an anxiety disorder are challenged and replaced with more realistic and positive thoughts.

This process takes place in three steps.

Recognizing your negative thoughts

People with anxiety disorders often see the specific situation as being more dangerous than it actually is. For instance, individuals with germ phobia might be afraid of touching the handle of a door as it might be viewed as a threat to their lives.

Sometimes it is difficult to recognize or identify your irrational fear and thoughts. A way of achieving this is to ask yourself what your thoughts were when you were feeling anxious or worrying.

Challenging or addressing your negative thoughts

This is the second step, and it involves working with the therapist to come up with strategies for dealing with the thought triggers for your anxiety. To do this, you will have to come up with evidence of your negative thought pattern. You can obtain evidence by analyzing your distorted beliefs and putting your negative prediction to the test. The therapist will help you to come up with assignments that will allow you to do this and let you experiment with the advantages and disadvantages of various strategies as well as help you to realistically gauge the probability of your imagined worst-case scenario coming to pass.

Replacing negative beliefs and thoughts with positive, realistic ones

After the irrational fears and negative feelings that trigger your anxiety are identified, they can be replaced or modified to more positive and realistic ones. This is done with the help of your therapist, who will act as your guide to calm you when certain situations trigger your anxiety and panic attacks.

To give you a better understanding of thought-challenging with CBT, let's consider this example.

John is afraid of taking public transportation because he is afraid of dying in a commercial-vehicle disaster. After John's therapist recognized these negative thoughts, he asked John to write down his thoughts to identify the various distortions and errors in his thinking. Below is what he was able to come up with:

1. Persistent/Challenging Negative Thought:

What if I am in an accident and die while using a public transport system?

Cognitive distortion:

Predicting the worst-case scenario.

More Realistic thought:

The public transport system is used by a lot of my friends and family members who are still alive, so it must not be as unsafe as I think.

2. Second Negative Thought:

Dying in a transport accident is a terrible way of dying.

Cognitive distortion:

Not thinking straight.

More realistic thought:

There are a lot of ways to die; transportation accidents are just one of them; besides, even private transport can get involved in a road accident.

3. Third Negative Thought:

I could even die by just riding on any kind of public transport.

Cognitive distortion:

Jumping to conclusions.

More realistic thought:

People don't just die by getting on to public transport.

The process of replacing distorted negative thoughts with more positive and realistic ones is not an easy task. These distorted thoughts have been programmed into the mind of the individual in question and have become a pattern of thinking that can last a lifetime. It takes a lot of hard work to break any habit. CBT works hand in hand with the homework given to you to make it easier to achieve this. CBT for anxiety also include:

• Lessons to help you recognize when you become anxious by its effect on your body.

• Coping strategies to help you relax when dealing with anxiety and panic attacks.

• Challenging your fears (both imagined and real)

One system of homework or assignment that facilitates recovery from mental disorders and worries is known as Exposure therapy.

Exposure Therapy

A situation in which triggers anxiety is often unpleasant and people with anxiety disorders do whatever they can to avoid it as much as possible. For instance, if you are afraid of heights or plane flights, you will do whatever is within your power to avoid getting into any situation involving them. For those with the phobia of public speaking, they might even go as far as avoiding speaking at their best friend's wedding. In as much as these situations can be unpleasant, they can be an important aspect of living, and avoiding them takes away the chance for you to overcome them. And the more you avoid them, the stronger they become.

Exposure therapy aims to bring you into contact with those fearful situations and objects. It works on the theory that with repeated exposure, you will become accustomed to these situations and become more in control of your anxiety and panic attacks. There are two ways of achieving exposure therapy.

- Being asked by your therapist to imagine these scary objects and situations in sessions.

- Facing these situations in real life, so you can apply what you have gained in therapy sessions.

Systematic Desensitization

Sometimes facing your fears right away can lead to devastating results. Exposure therapy usually begins with situations that trigger mild anxieties and worries. It gradually works up from there toward more dangerous situations. This process of gradual exposure is known as systemic desensitization. It lets you build up tolerance and confidence for mastering control of your anxiety.

For instance, the process of systemic desensitization for fear of plane flights involves:

Step 1: staring at a picture of airplanes

Step 2: watching videos of aircraft in flight

Step 3: seeing a real plane taking off

Step 4: booking a flight ticket

Step 5: driving to the airport

Step 6: checking in for your flight

Step 7: getting on your plane

Step 8: taking the flight

Systematic Desensitization Occurs in Three-Phases

Learning relaxation skills

The first step involved in systemic desensitization is learning how to relax in those situations that trigger your anxiety. With the help of your therapist, you will be taught relaxing techniques like muscle relaxation and deep breathing. These techniques will be practiced both at home and in therapy sessions. These techniques will help reduce the physical symptoms (sweating, hyperventilating, and trembling) of your anxiety attacks as you confront your fears.

Create a step by step list

A list of those situations that trigger your anxiety will be created to help guide you toward your goal. A list of actions to overcome the fear of each situation will be made to provide a guide and strategy. Every step must be specific with a realistic and measurable objective.

Making anxiety therapy work for you

Anxiety needs a lot of hard thinking and time as well as commitment. Treatment with CBT requires you to face your fears, and sometimes you may end up feeling worse before getting better. But whatever the case may be, sticking to the treatment plan and listening to your therapist is paramount to your success.

Chapter Five: Dealing with Depression: CBT Tips to Feel Better Instantly

Life can get complicated at times, so it is perfectly normal to feel down once in a while. Feeling like everything is against you from time to time is an all too common feeling and one of the characteristics of being human, especially in today's society. According to a report by the Anxiety and Depression Association of America, 14.6 million people live with major depressive disorders.

Today many people work for longer hours only to receive the same pay; some have to deal with too many bills or personal relationship problems. Other people are dealing with addiction problems such as alcohol and drugs. Everyone has many issues, and not everyone is a hundred percent all the time. But when your feelings are at their lowest all the time to the extent that they impair your everyday life, or you experience feelings of despair that just won't go away; you might be experiencing depression. Depression is a dark and lonely place that can make day-to-day functioning a challenge. Some days you are overwhelmed, and the only solace you can find is in alcohol and drugs.

If you are in this place right now and feel like no one is coming to save you, the good news is you can help yourself. CBT had been proven to be extremely useful in helping people living with depression.

Types of Depression

Many people experience various kinds of depression. These different types of depression can be experienced either together or in addition to an addiction problem; whatever the case might be, CBT is useful for treating many kinds of depression.

Major Depression

This type of depression occurs when you have experienced five or more symptoms of depression for at least two weeks. Major depression is often weakening and interferes with a proper daily function such as work, sleep, eating, and studying. You can experience episodes of major depression a few times in your life. Sometimes they can happen as a result of traumatic events such as the death of a loved one or the breakdown of a relationship.

Bipolar Disorder

People with the type of depression experience symptoms of shifting moods. It involves a cycle which goes from feelings of mild to intense happiness (euphoria) to episodes of extremely overwhelming depression.

Persistent Depressive Disorder (PDD)

This type of depression was previously referred to as dysthymia. PDD is usually a less severe type of major depression, although its symptoms are often similar to that of major depression. People with this type of depression typically experience it for a least two years.

Some symptoms of PDD include stress, irritability, and the general lack of the ability to enjoy life.

Signs and Symptoms of Depression

Often, people are worried that they are experiencing depression. Because many people often feel sad from time to time, it is essential to be able to tell the difference between experiencing depression and just dealing with a short period of sadness. Depression can usually be identified by loss of interest in life in general and the inability to carry out daily functions effectively. The following signs and symptoms generally define depression.

- Lose interest in things you used to enjoy

- Constant feelings of helplessness and hopelessness

- Unexplainable tiredness

- Inability to concentrate, even when the task is easy

- Change in appetites — either eating more or less

- Failure to think anything positive.

- Aggressiveness, irritability, and short temper

- Drinking more alcohol than usual

- Engaging in reckless behavior

- Excessive use of prescribed or illegal drugs.

- Feelings of guilt and worthlessness, or self-loathing

Experiencing all of these symptoms is a sure sign of depression, and one of the best options of treatment available today for you is CBT.

Cognitive Behavioral Therapy for Depression

Having good knowledge of the symptoms and being able to identify them is the first step to recovery; many people sometimes find it difficult to identify signs of depression.

The next step is knowing about the various effective treatment plans for depression.

Cognitive-behavioral therapy is a psychotherapy that helps in modifying thought patterns to change negative moods and behaviors into more positive ones.

Treatment of depression with CBT applies both cognitive and behavioral therapy. With the help of your therapist, you will be able to identify those negative thought patterns that trigger inappropriate behavioral responses to certain situations.

The treatment plan follows a structured pattern to guide you in coming up with strategies to help you deal with those situations that serve as triggers. These strategies help in managing or eliminating your depression. CBT aims to improve your present state of mind rather than dealing with the past.

As with most mental illness, treatment of depression with CBT is a difficult task. To help you, below are some guidelines to help prepare you for treatment.

Therapy

Since CBT is a goal-oriented method of treatment, it doesn't require as much time as other methods of treatment. Therapy sessions could be once every week and could last for 30 or 60 minutes

The first set of therapy sessions will be used to determine if you need the treatment or are a fit for it and to know if you are comfortable with the procedure.

Although the main focus of CBT treatment is on your present life, your therapist will need some understanding of your past, so you will be asked some questions about your history and background.

You will make the final decision on the changes you want. Also, you and the therapist will make the decisions on what you want to be discussed each day.

CBT Treatment for Depression

With the help of your therapist, every one of your problems will be broken down into smaller, more manageable parts. Each part will be taken separately and solved following the laid-out plan. You will be asked to keep a record of your thoughts, emotions, feelings, and behavioral patterns to be able to identify and modify them.

The record will help your therapist determine how those thoughts, feelings, and emotions are affecting you, and to see which of them may be unrealistic and distorted. You will then work with your therapist to come up with strategies on how to cope with them, and then to gradually modify them.

You and your therapist will work together to come up with homework or assignments that will help you to practice and apply things learned in the sessions.

Further therapy sessions will provide the opportunity to see how much progress has been made so far from the previous session and to see how well the last assignment was accomplished.

Unlike other kinds of psychotherapy, CBT treatment requires a good relationship between the therapist and you, so every decision made will be made together. In order words during the procedure, no choice will be forced on you by the therapist.

Even after you are done with the sessions, you can continue to apply these strategies; this enables you to be healthy for as many years as possible.

How CBT Works for Depression

One of the unique features of CBT is that it requires less time, taking as little as 6 to 20 sessions.

During every session, you and your therapist will identify situations that contribute to depression and try to tackle those patterns of thought. The journal or diary used to keep records helps your therapist to break down those thought reactions and patterns into different groups such as:

All-or-nothing thinking, where your view of the world is black and white, or

Generalization of everything, which refers to using the result of an event to judge other events.

Automatic negative thought patterns, when certain circumstances trigger a series of negative thoughts which has become habitual.

Not believing the positive, when you always consider the prospect of any positive experience as something that can't possibly happen.

Minimizing or maximizing the importance of certain events, when the critical or non-critical natures of certain situations are distorted.

Blowing things out of proportion, wherein you always think everything that happens is as a result of what you have done or said, or that behaviors and activities of people are because of you.

Focusing on an adverse event, when you always tend to dwell on an adverse fact so that your view of reality is distorted

Keeping a record of things also helps you:

• To analyze yourself to come up with appropriate ways of responding to situations.

• To know how to talk to yourself in realistic ways.

• To be able to analyze your emotions and situations accurately.

• To be able to come up with appropriate responses to specific events.

Applying these methods and techniques help you to gain a balance with your mind and body.

How CBT Helps Depression

Depression has become one of the most widespread mental health problems experienced by both old and young, and the debilitating effects of depression can't be ignored. This mental health problem goes beyond affecting just your life; it also affects friends and family members. Depression is a common and severe illness that negatively impacts the lives of those close to the sufferer; family and friends as well as coworkers and employers.

Depression significantly affects the proper functioning of society as a whole. For example, when depression keeps you from carrying out your proper function at works and impacts your financial life, you won't suffer alone as the effects extend toward your family, employer, and anyone who tends to gain from you financially.

What Kinds of Depression does CBT Treat?

CBT is useful for treating people with moderate depression, and it can be used as a treatment plan without the need for medication. For those with significant depression, CBT works best when used together with drugs.

Just as depression affects both old and young, CBT is also effective in treating both, and it goes a long way toward reducing risks of relapse. The coping strategies and cognitive modifications derived from CBT treatment give you long term skills to deal with a lot of demanding situations. So, CBT is a useful tool to stay mentally healthy and free from depression for a long time.

For CBT to work, you must

· Be motivated to change your current situation

· Be capable of introspection

· Have the ability to control your reaction to things happening around you.

How the Components of CBT Work on Depression

CBT is a psychotherapy that has two components: the cognitive part and the behavioral part.

The cognitive part helps to identify those unrealistic negative thoughts that lead to negative behaviors and emotions. It also helps you to understand those beliefs that you have developed over time and what triggered their development. This is an essential part of CBT treatment for depression.

The behavioral part helps you to deal with treatment and modification of the various responses to and behavior in certain situations. With the help of your therapist, you will analyze your daily activities and their effects on your moods.

CBT goes beyond the therapy sessions as you will be given various assignments to practice everything gained in the therapy sessions.

Depression and Addiction

One of the most common associations of depression is an addiction. People who are depressed are at high risk for abuse and dependence on substances that numb those painful feelings. Sometimes abuse of specific substances such as alcohol depresses your central nervous system. Abuse of alcohol could, therefore, serve to induce depression. Twenty percent of Americans who suffer from anxiety and mood disorders like depression are addicted to alcohol or other substances.

Based on the statistics, it seems a fact that depression and addiction are closely related, and each of the condition tends to amplify the other. It is essential to seek help if you are experiencing both states at once.

The interrelationship between depression and addiction is referred to as dual diagnosis. People with both conditions often see life as being extraordinarily lonely and weakening, because both conditions serve to make the other worse.

Triggers of Depression and Addiction

Most times, it isn't easy to determine whether depression or addiction started first. But based on years of study and research, some triggers of both conditions have been identified.

Both conditions seem to affect the same area of the brain. This area of the brain is also responsible for how we respond to stress.

Genetic factors also have a significant role to play in substance abuse and depression. Certain people, due to their DNA, are more prone to depression and addiction.

Early development problems affect the mix; people who had mental health problems while growing up are more likely to become addicts, and people who had issues with substance abuse at an early age are more susceptible to mental health problems in the future.

Things to Remember About CBT

New experiences can be challenging, especially the life-changing ones, and going for CBT as your choice of treatment means you are going for something challenging. Doubts and worries about it working are perfectly normal, but you will be required to put in the work, and with the help of your therapist, a good result is guaranteed. Before going for CBT treatment for depression, it is essential to know the following.

CBT explores those painful experiences and feelings that you always want to avoid, so you might end up facing these situations.

Achieving your goal of having good mental health is totally up to you. Your therapist can only serve as a guide to encourage you, but in the end, you end up doing all the hard work.

You will need to desire to be well, so you will have to push yourself even when you don't feel like it.

If you want to get well, confronting those situations you normally try to avoid is going to be necessary.

Chapter Six: Workplace CBT: Ways to Beat Stress at Work

Concern for work-related stress is growing worldwide, and if you are worried about the amount of pressure you go through at work, then you are not alone. It is an issue that goes beyond just affecting the health of employees, impacting the efficiency and productivity of affected organizations. Many companies demand maximum commitment and have an enormous workload to be dealt with by their employees.

Many events serve as triggers for work-related stress. For instance, you might be overwhelmed with the workload you face, and the pressure of the demand it places on you to get it done might be impossible to deal with. Many jobs can require an absurd number of hours that you feel it isn't worth it. Different situations at work that can be a source of stress include:

· Work conflicts between colleagues or your employers

· Dealing with constant change at work

· Fear of losing your job or being demoted

Work pressure affects various individuals differently; not everyone has the same view of working. What you may view as challenging another person may see as extremely stressful while some other person might not feel it is much of either a challenge or stressful. This is because not everyone has the same psychological constitution. People have different experiences, and as a result don't always have the same views on certain things. But the fact remains that everyone feels stress when dealing with specific challenges in life.

Individual events can increase or decrease stress. Symptoms of work-related stress could be either physiological or psychological. Many individuals are searching for ways to reduce work-related stress. One method of dealing with stress that has proven useful in treating many aspects of psychological problems is cognitive-behavioral therapy (CBT). CBT helps people going through work-related stress to find a new perspective on their situation. It helps them manage pressure while also decreasing the effects of psychological and emotional stress. CBT also teaches new strategies to help them feel at ease and have more confidence in the face of any work challenge.

Identifying some specific symptoms of stress can be tricky, while other signs might be mild and manifest in smaller details. Whatever the case might be, you must recognize the symptoms of work-related stress to know when to seek help.

Symptoms of Work-Related Stress

Many studies have been carried out on the symptoms and effects of work-related stress

Symptoms such as upset stomach, headache, sleep problems, and relationship problems with friends and family are well-known signs of stress at work.

Symptoms of work-related stress are divided into three categories: psychological, physical, and behavioral.

Psychological Symptoms

Depression

Dermatological disorders

Discouragement

Pessimism

Anxiety

Irritability

Cognitive difficulties

Some of these psychological problems are easy to identify. The physical effects of stress at work, on the other hand, are not so easy to recognize because they are associated with different problems and illness. Work-related stress may precipitate chronic disease. A study published by the Journal of Occupational and Environmental Mental Health revealed that the cost of health care is over 50 percent more for workers who experience high levels of work-related stress.

Cardiovascular Disease

Some jobs are extremely demanding and constantly changing, and do not give employees control over what is happening; these jobs increase the risk of cardiovascular disease.

Musculoskeletal disorders: specific jobs requiring certain forms of physical activity increase the risk for the development of diseases of the upper limbs and back.

Workplace injury: Stressful working conditions usually interfere with proper safety practices, increasing the risk of work-related injuries.

Suicide, cancer, ulcers, and immune function: According to some research, there exist some definite relationships between work-related stress and health problems such as these.

Physical Symptoms

Headaches

Fatigue

Sleeping difficulties, such as insomnia

Muscular tension

Heart palpitations

Gastrointestinal upsets, such as diarrhea or constipation

High blood pressure

Loss of appetite

Poor job performance

Behavioral Symptoms

Aggression

An increase in sick days or absenteeism

A drop in work performance

Diminished creativity and initiative

Mood swings and irritability

Lower tolerance of frustration and impatience

Disinterest

Interpersonal relationships problems

Isolation.

Short attention span

Procrastination

Increased use of alcohol and drugs

Triggers of Work-Related Stress

These are some factors that act as facilitators for pressure at the workplace:

Bad management

High-performance demands

Work environment and surrounding

Lack of proper support

Changes in management

Trauma

Role conflicts

Causes of Work-Related Stress

These factors are the major factors responsible for fear in the workplace.

High workloads

Long hours

Short deadlines

Job insecurity

Changes within the organization

Insufficient skills to get the job done

Boring work

Poor relationships with colleagues and employers

Over-supervision

Changes to duties

Lack of proper resources

Bad working environment

Not enough promotion opportunities

Lack of equipment

Discrimination

Harassment

Random events in the workplace, such as workplace deaths.

How Does CBT for Stress Work?

CBT treatment for work-related stress helps in providing understanding about the effects of specific thinking patterns on our behavior and how those can raise your stress level. Also, while it helps you with identifying these thinking patterns, it helps you in creating new thinking patterns that change your behavior and response. It also serves to help boost your confidence and ability to cope with certain stressful and challenging situations

After going through cognitive behavioral therapy, you will be able to control your behaviors better and handle stressful situations with ease. You will also know how to prevent some situations at work being stressful at all.

Therapy

For the first CBT therapy session you will be asked various questions by your therapist, so the amount of help and the approach to take in handling the challenges you face can be determined. This also helps your therapist to come up with an appropriate plan to achieve the goal.

Subsequent sessions will be used to determine and identify the situations that act as triggers for you. This is done through a thoughtful discussion with your therapist. This helps you in knowing and seeing these triggers from a new angle. Also, you will learn new ways of thinking, handling, and coping with those stressful situations.

Various assignments will be given to you to help you put to work everything learned in the CBT sessions and to see how much your capacity for dealing with stressful situations at work has improved. There is no easy way out of work-related stress; you will have to put in a lot of hard work because these stressful situations are all stressful for a reason, and until you figure out how to deal with these reasons, your condition won't change.

Some practical tips from CBT which can help you in dealing with work-related stress include:

Learn to prioritize

Sometimes having too much on to do doesn't inspire you to do more; it may just add to an already stressful situation. You might end up feeling overwhelmed and feel like everything is out of your control.

You don't have to do everything. Learning how to prioritize is an essential aspect of working for some organizations — taking your time to prioritize makes things run more smoothly. You might even find out you have time for many other things that can serve to reduce work stress. You can make a list of the most important things to do as opposed to tasks that are not essential. This gives you a higher chance of being in control and taking your jobs one at a time from the most important to the least important.

Monitoring your mood

This is an essential tip of work stress management from CBT. Your therapist will help you in coming up with ways to control how certain situations and events at work affect your mood. In other words, it helps you to process how you feel towards particular circumstances while aiding you to see how some behavioral patterns affect your mood in specific ways.

When you find yourself focusing on particular thoughts like how much workload you face and worrying about yet-to-come situations at work, your mood-monitoring skill might come handy. You can get a journal to record your moods.

Record the stressful situation

Record how you feel at the time of the trigger situation or any time you think about the situation. You can rate each feeling based on how overwhelming it was.

Record everything you were thinking at the time of the situation. It is essential to get every single thought down, so you can tell how each one affects your feelings.

When you are done recording everything, you can put your journal away. After a few days, you can revisit your journal to go through what you had put down.

This way of recording everything that goes on with your thoughts and feelings is a great way to teach yourself to see your emotions from another angle. You are more likely to notice just how distorted your views and attitudes were at the time of the situation when you visit your journal a few days later. This gives you the chance to be better able to recognize which thoughts need modifying and changing so next time you can better deal with and respond to the situation.

From CBT, you can learn how these negative thought patterns (cognitive distortions) affect your mental health and how we can come up with strategies for dealing with them.

Focus on the things you can control and develop a positive balance

In situations where we feel overwhelmed and consumed by our workload, we often tend to focus only on things we can control. This often ends up raising our overall stress level and exhausting our minds, using up energy that we could be using to achieve something better.

In times like this, CBT lessons for reframing our minds and thinking can help us feel in control.

Positive reframing is different from just positive thinking. Positive reframing helps you come up with new ways and strategies using the facts available to look at things in a more realistic manner. Instead of just coming up with positive thinking towards the stressful situation or task, positive reframing comes up with an alternative way of solving or coping with the situation. For instance, you could be in charge of planning the end-of-year party for your organization/company and, at the same time, managing your already busy schedule. Positive reframing could help you come up with an alternative way by helping you see the importance of delegating instead of just handling everything yourself

Situation: a big upcoming event that requires a lot of details and inputs. Thoughts: *This is so much work for me; I don't think this work can be just for me. Doing this alone could result in a big disaster.*

Emotions: irritable, depressed, and anxious.

Behaviors: avoid everything to do with it. Avoid doing the project, procrastinate, and leave out important details of the project.

Alternative thought: *Although this is a lot of work for one person, I have always been good at accomplishing jobs like this, and for me to be assigned this task means my boss thinks highly of me, which means I can do this and must not disappoint. I'm going to do every bit of research required for this task; this could be my chance to show my stuff. I can always ask for help if I'm stuck, so I will carry it out diligently to see how far I can go on my one for now.*

Look for satisfaction and meaning in your work

Sometimes we may end up feeling dissatisfied and bored with the constant stream of work. This is a significant cause and source of stress for many people and can affect your mental and physical health. Some of us have always dreamed of the perfect position, career, or job. One of the greatest motivations and source of drive that get many people going is being passionate about their work. Once this is lost, you are going to feel dissatisfied.

Maybe you are not in your dream job, but you can still find purpose in being there. With ambition, you can even learn to develop a passion for the job. Even in some extremely unimportant tasks, you can learn to find meaning in the little contributions you make. All you have to do is focus your attention on the aspects of the job you love and enjoy. Even if it doesn't seem like much to others, you might find out that with time you might get the promotion you seek.

At other times, when you feel helpless and uncertain, when the level of your stress is over the roof, some of the tips below can be of help to you:

Speak to your employer about workplace stressors

Your employers are aware that happy employees of sound mind are more efficient and productive, so they will always try their best to tackle work stress to get the best out of the employees. So, it is important to let your employers know of those stressors that makes it impossible to carry out your job effectively

Get a clear description of your job

If you don't have a good understanding of your responsibilities and duties concerning a task, you might find the task very difficult. This can increase the level of your work-related stress. You can always ask for clarification on a mission, so you know what you are doing.

· You can ask for a transfer into another department to escape the toxic environment.

· If you are tired, bored, or stressed out by the same old task, you can ask for something new.

Chapter Seven: Intrusive Thoughts: Acknowledging and Eliminating Them with CBT

Sometimes you might experience specific thoughts that pop into your head from nowhere. Maybe you're just going about one of your daily activities, and suddenly your mind comes up with a bizarre thought or crazy image, leaving you wondering where it came from. Most of the time, the idea could be harmless such as doing something stupid and socially crazy in public. Sometimes it could be a thought that could do more harm than good, or something that you could never dream of doing, like pushing someone down a flight of stairs.

The good news is you are not the only one experiencing strange and bizarre thoughts popping into the mind at random times.

What are Intrusive Thoughts?

Intrusive thoughts are thoughts that come into our consciousness without any prompting or warning. The contents of these thoughts are sometimes usually unacceptable to the general population, as they are disturbing and alarming or just weird. When these thought gets stuck in our head for some reason, they can lead to severe distress.

In some situations, where such thoughts occur frequently, they may start interfering with our everyday life. These thoughts could be of behaviors that are violent in nature, sexual, and other disturbing fantasies that are unacceptable to you.

It is essential to know these thoughts are nothing more than thoughts and have no meaning whatsoever, so the power they have over you will be only that which you give to them. When you focus more attention than is necessary on these thoughts and get worried over them, feeling ashamed and becoming disturbed by them, then you could be experiencing a mental disorder.

When you know that these thoughts are nothing but thoughts, and you have no obligation to do as they suggest, intrusive thoughts can't be harmful.

What Causes Intrusive Thoughts and Are They Normal?

The cause of intrusive thoughts hasn't been determined for sure, but some psychologists have published some theories. Lynn Somerstien proposed that maybe the reason for these thoughts surfacing is because the person is going through something difficult. This situation could be interpersonal problems, work stress, parent and parenting problems, or something which the person is trying to keep under wraps. Unfortunately, instead of the thoughts of these problems staying buried, they find an alternative way of manifesting.

Another psychologist who has proposed another theory is Dr. Hannah Reese. She suggested that the manifestations of these thoughts are a result of our failure to act in the way they suggest, because while you will never do as these thoughts suggest, your brain goes right on coming up with some of the most bizarre things it can imagine.

This brings us to the question of why our brain keeps coming up with such thoughts.

Dr. Sally Winston and Dr. Martin Seif came up with an outstanding description of what they believe causes intrusive thoughts. They believe that our brains create what they call "junk thoughts" and these are part of the debris that floats around in our stream of consciousness. Thoughts like these are meaningless, and if we avoid them and ignore them, they just disappear.

The argument of where these intrusive thoughts come from remains a mystery, but the fact remains that in some cases people dwell too much on them, and the more they try to avoid it, the more they think about it.

In some other cases, these thoughts come as a result of an underlying mental health problem or a brain problem such as

Post-traumatic stress disorder (PTSD)

Obsessive-Compulsive Disorder (OCD)

Brain injury

Parkinson's disease

Dementia

It is essential to notice changes or symptoms in your mental health because they are not to be taken lightly. Some early signs of mental health problems include

Changes in thought patterns

Thoughts of disturbing imagery

Obsessive thoughts

For instance, if someone tells you to avoid thinking of a green whale, even though you are allowed to think about anything else in the world, don't think about a green whale, it can be hard avoiding the thought of a green whale, especially for a long time. In time you will find that your mind will slip up and the image of a green whale will come to mind.

In a healthy mental state, it is easy for you to monitor and keep track of your many thoughts, and even when the little random reflections come up, it is easy to let them slip away.

In situations where you find it difficult to let go of these intrusive thoughts but instead you keep on focusing on them more and more frequently; then it is essential to seek help.

Intrusive Thoughts and Other Mental Health Disorders

Some of the mental disorders most associated with intrusive thoughts include:

Anxiety;

OCD;

Depression;

TSD;

Bipolar Disorder;

ADHD

Intrusive thoughts popping up in our minds is normal; everyone experiences it. However, in cases where they come up more frequently and lead to significant distress you may have one of the associated mental disorders mentioned above.

Intrusive Thoughts and OCD

One of the significant easily recognized symptoms of OCD is frequent intrusive thoughts, experienced by almost all persons diagnosed with OCD.

According to Dr. Robert L. Leahy, these thoughts are often evaluated negatively: You might end up thinking that there is something wrong with you because these thoughts which you should not be thinking of keep popping into your mind. So, the only way you see of controlling them is to pay close attention to them, to monitor and prevent them from coming up.

People with OCD who experience these compulsive intrusive thoughts react in a certain way to these thoughts, leading to more severe problems. The frequency of these thoughts only escalates with more attention paid to it, leading to them becoming an obsession. This obsession results in repeated behaviors carried out in order to avoid the recurrence of these thoughts.

Some examples of intrusive thoughts relating to OCD include worrying about shutting the windows, having the key to the door, and worrying about batteries on surfaces. Someone with OCD may develop the habit of cleaning surfaces multiple times or avoid touching the handle of the door or rechecking repeatedly to be sure they have their keys with them. These compulsions often affect the quality of life of the individual and interfere with a healthy daily life.

Intrusive Thoughts and Depression

People suffering from depression are also prone to having intrusive thoughts. Frequent depressive intrusive thoughts can also cause depression. When too much attention is placed on specific negative and depressive intrusive thoughts that frequently occur (rumination), it can lead to severe depression. You might return time and time again trying to address these thoughts, but instead of solving the issue you only end up making it worse.

Some examples of intrusive thought within depression include:

· Placing too much focus on negatives and always expecting worst-case scenarios.

· Placing too much emphasis on a specific horrible event and using it as a reference to other similar events in the future.

· Over-analyzing things in your head (over-thinking)

· Always assuming you know what others are thinking

· Accepting the worst-case scenario as the only possible outcome of a particular situation

· Exaggeration of a certain event

· Taking responsibility for things that you can't control

Thoughts like these can cloud your mind and make it impossible for you to see things as they actually are. Instead of seeing most of what goes on in your head as just thoughts, you end up believing them, taking every analysis as real and not being objective in your conclusions.

Intrusive Thoughts and Anxiety

In cases of OCD, the person involved tends to experience intense graphic, violent, and unacceptable intrusive thoughts, while people with anxiety often feel like they are drowning in many unwanted thoughts of less intensity than those of OCD sufferers.

In the case of a Generalized Anxiety Disorder (GAD), patients may experience uncontrollable worry about the safety of a loved one. Certain people with an anxiety disorder relating to fear of social situations (social phobia) might find it challenging to get over memories of making a mistake or saying something they shouldn't have said.

Usually when someone with an anxiety disorder experiences an intrusive thought, they will end up making the worst decision concerning the negative feeling. They often spend more time than is necessary obsessing over the thought, all in the name of trying to get it out of their mind. When they spend additional time on an intrusive thought, they often give it power over them, losing control over their minds as a result.

Intrusive Thoughts and PTSD

Another mental health problem strictly associated with intrusive thoughts is post-traumatic stress disorder (PTSD). In the case of PTSD, the intrusive thoughts are related to a particularly traumatic event that has already taken place, and could even involve flashbacks to it.

People with PTSD have difficulty forgetting what has happened in the past to them; as a result, the symptoms of PTSD make them deal with the past over and over again. They experience flashbacks from time to time in the form of nightmares and intrusive thoughts. In episodes of PTSD, the state of the body is similar to that of the previous situation. As a result, the person is on high alert due to floods of the "fight and flight" hormones and other hormones to the brain.

Intrusive Thoughts and ADHD

A primary symptom of ADHD is intrusive thoughts. People with ADHD often find it difficult to pay attention, even in the most conducive of environments. Difficulty concentrating is a general feature of this mental health condition, and one cause for this is frequent disturbing intrusive thoughts. People with ADHD experience a higher degree of intrusive thoughts than those with OCD even though the disorders are similar.

Cognitive Behavioral Therapy (CBT) Treatment for Intrusive Thoughts

CBT is one of the most effective treatment options for intrusive thoughts. Since intrusive thought has to do with how these how random thoughts pop into the mind and influence behavior, CBT is nothing short of a perfect choice of treatment for this mental health condition. CBT is used alone or in combination with other options of treatment depending on the severity of the condition.

CBT helps in creating and coming up with management strategies for dealing with harmful and undesired thoughts and behaviors. Through CBT, you will be able to learn how to come up with healthier ways of ignoring intrusive thoughts.

Acceptance and Commitment Therapy (ACT)

Acceptance and Commitment Therapy is a subtype of CBT that teaches you to accept your feelings and thoughts instead of engaging in a battle to avoid them. ACT shows you how to be mindful while coming up with alternative ways of thinking. It teaches people with these intrusive thoughts to accept these thoughts as normal but not to dwell on them as they are just one of many thoughts that can be ignored. The six principles of ACT are:

Cognitive Diffusion: You learn to give little weight to negative thoughts, emotions, and images.

Acceptance: learning how to let those intrusive thoughts run through your mind without feeling distressed

Contact the present moment: learning to focus on the current instead of overthinking the past or the future, and learning to accept things going on around you.

Observing yourself: Being aware or conscious of your being.

Values: identifying those important values your life is based on, those things you find the most important.

Committed action: Assigning goals depending on your values and the things you are fighting for.

These six principles help in treating and healing you while creating a forward-thinking mind.

Exposure and Response Prevention (ERP)

Another aspect of CBT that has proven effective in helping people with OCD achieve stable mental health is Exposure and Response Prevention (ERP). In this method of therapy, you are exposed to the situations and events that act as triggers of your fear, and learn how to deal with them better.

ERP aims to show you that you are capable of challenging those fears, so you will come to realize how irrational they are. Those intrusive thoughts might remain, but with the help of this therapy, they become nothing more than a negligible nuisance that you don't pay much mind.

Self-Help: Managing Intrusive Thoughts with CBT

This method is used in addition to other CBT methods to lessen the symptoms of intrusive thoughts and give you a better quality of life when faced with intrusive thoughts.

According to Seif and Winston (2018), there are seven steps that can help you in changing your attitude towards intrusive thoughts and in getting over them.

- Give these thoughts a label, such as "intrusive thoughts".

- Know that you don't have control over these thoughts, they are automatic.

- Do not push the thoughts away, accept them.

- Float, and pass away the time.

- There is no need for a rush. Give yourself time, remember that less is more.

- The thoughts will come again, expect them.

- You can allow the anxiety to be present, but do not stop what you were doing before the intrusive thought.

Also, Seif and Winston put up some warning signs against these thoughts.

- Engage the thoughts in the best way you can.

- Do not keep the thoughts in your mind.

- Find the meaning of the thought.

- Observe to see if this is effective in getting rid of the thoughts.

The North Point Recovery center, which is an organization that helps people dealing with various disorders and substance abuse, came up with five tips to help people in challenging their intrusive thoughts.

- Take a more in-depth view of why intrusive thoughts bother you.

- Don't block your mind. Allow the thoughts in and move on from them.

- Don't be triggered by thoughts; they are just thoughts, don't give them more power than they have.

- Don't react emotionally to intrusive thoughts.

- Trying to align your behaviors with your obsession won't help in the long run.

Chapter Eight: Mindfulness and the CBT Connection

With the increase in popularity of both CBT and MCBT treatments for various mental health issues, many questions have been asked about what the two offer the world of psychology. The more popular Cognitive Behavioral Therapy (CBT) has gained a broad audience for its practical and goal-oriented method of dealing with many illnesses. The relative newcomer, MBCT, still has a long way to go to attain the popularity of CBT.

Because of the many values and similarities, it is often difficult to tell the difference between the two methods.

To give you a full picture of the basic principles and what both methods of psychotherapy are about, we are going to go through a summary of both approaches.

Cognitive-Behavioral Therapy (CBT)

For this comparison, we are going to go through a short overview of CBT, which we've already seen in chapter one.

As the name implies, Cognitive Behavioral Therapy is a type of psychotherapy that applies two components of treatment: the Cognitive and the Behavioral components. By using these two parts in its goal-oriented treatment plans, it has been successful in treating various mental health problems such as anxiety disorders, PTSD, depression, schizophrenia, and OCD, among others.

Cognitive component: the cognitive component is the part of CBT responsible for recognizing those distorted thoughts and modifying them into more realistic ones. You might be experiencing specific thoughts and feelings that serve to make you have some distorted beliefs. When you act on this unrealistic belief it often leads to specific behaviors that can interfere with healthy living and many aspects of life, such as the relationship between family members, romantic relationships, academics, and work.

For instance, when someone suffers from low self-esteem, they might be dealing with some distorted (negative) thoughts about their capabilities and appearance. This could result in negative thinking patterns that might tend to keep them away from social events or give up specific opportunities that involve dealing with or exposure to people.

The cognitive component of CBT addresses the actions of modifying and changing these destructive thoughts. With the help of your therapist, you will be able to identify those distorted thought patterns and beliefs. This stage of CBT is referred to as "functional analysis". This stage is vital, so you can move forward with determining how these thoughts affect your behavior.

Behavioral Component

This part of CBT deals with the resulting behaviors due to the distorted negative thoughts. These behaviors are the end products of false and unrealistic beliefs due to negative thinking patterns. With the help of the behavioral component, you will be shown a new skill or strategy for coping with these behaviors, which can be applied to actual life situations.

In most cases, a change of behavior is accomplished in many gradual steps.

An instance of CBT in action would be when you are supposed to go out with a friend and he turns you down, saying he is busy. You might end up thinking that he hates you and so wants to stay away from you, especially if this happens repeatedly. This leads to more negative thinking, leading to you doubt and question your worth. You might end up feeling anxious and paranoid, so the next time you have an outing you end up using your previous experience to judge it.

In CBT treatment, you will be taught how to recognize and identify these negative thoughts. Instead of believing them, you will be shown how to look at an alternative pattern of analyzing the situation. You will learn how to question all your negative assumptions. You will be asked to consider the other previous outing you have had with your friends or someone. After going through all these, maybe you will be shown that perhaps all these thoughts are just in your head, and perhaps the friend who turned you down is indeed just busy.

What is Mindfulness?

It is a widespread belief that our reality is defined by the way we think. It's also believed that this reality can be influenced by improving the quality of one's thoughts. Every thought and feeling you experience shapes the nature of your reality. Mindfulness-based cognitive therapy, or MBCT, helps you recognize and understand the tone of your thoughts and feelings and to create new, healthy habits.

MBCT effectively combines cognitive therapy along with techniques of mindfulness to help an individual deal with issues like anxiety, depression, or any other behavioral problems. It mainly helps lessen your worries, stress, and fears by enabling you to control your emotions.

The ability to be aware of the thoughts popping into your head without getting carried away by them is known as mindfulness. The mind tends to wander, and as you try to concentrate on the task at hand, you might notice other thoughts creeping in. Mindfulness enables you to control your mind by using techniques that encourage you to take stock of your thoughts and decide whether you want to respond to them or not.

Mindfulness psychotherapy is designed to make you focus your awareness on the present moment. It helps you to analyze your feelings, thoughts, and bodily sensations calmly.

The foundation of mindfulness is based on an ancient technique used by Buddhist and specific Eastern spiritual teachings, and is designed to help people in attaining awareness of their body, feelings, and mind in order for them to gain self-actualization.

Mindfulness was developed in the 1970s by Dr. Jon Kabat-Zinn, who was the head of a stress reduction clinic at the University of Massachusetts. It was used as a psychological tool in the control of stress, anxiety, and chronic pain. It was researched and used in the treatment of depression in the 1990s. Today MBCT has been scientifically researched and is recognized by many of the world's leading psychologists, doctors, and scientists.

Mindfulness has been useful in helping people in dealing with the "auto-pilot" life we live in today's modern world. It helps us in always staying conscious of the present. This is important when dealing with mental illness like depression. Allowing the subconscious to rule over our lives gives room for a specific psychological condition like anxiety to come into our life. Getting distracted can leave us open to being taken over by particular challenges. If this happens, our reaction to it is bound to be automatic, and we can overreact and go off the rails. When we are ever-conscious of our present and aware of everything around us, we have a higher chance of responding calmly to individual challenges, events, or situations.

With the help of mindfulness, we think carefully, considering every option available before responding or acting. So, before acting, we consciously acknowledge the people, environment, and everything that will be affected by our action.

What is MBCT?

Mindfulness-Based Cognitive Therapy (MBCT) is a combination of various aspects of Cognitive Behavioral Therapy and mindfulness.

According to the two psychologists, Philip Barnard and Jon Teasdale, the human mind is made of two different modes, the "*being*" mode and "*doing*" mode. They described the "doing" mode as goal-oriented, and it is active when you come across a difference between how you want a thing to be and how the situation presents. The "being" mode, on the other hand, accepts situations the way they are without doing anything to change them. They further went on to say that the "being mode" is the one associated with long-lasting emotional changes. So, it was concluded that for cognitive therapy to be effective it will have to support not only cognitive awareness like CBT, but also the "being mode" of the mind. They believed that cognitive therapy could only be effective when used in combination with mindfulness.

A combined effort of psychiatrists Jon Kabat-Zinn, Zindel Segal, and Mark William helped in combining the various new ideas of cognitive therapy with the mindfulness-based stress reduction program of Kabat-Zinn. This led to the birth of MBCT.

The goal of MBCT is similar to that of CBT in that it helps you in maintaining a constant awareness of your reactions and thoughts. This enables you to notice any change that occurs due to negativity. But MBCT includes something extra in that it shows you how to become aware of time or moments when you are triggered by any negativity.

With this, you can better manage and control anxiety and stress by becoming more aware of what is happening at the present moment. So instead of putting so much attention on trying to understand your thoughts, with MBCT you accept it for what it is without any judgment; you just let it go through your mind without paying much attention or attaching much meaning to it.

More awareness of the present moment means it is less likely you will be caught off guard by any trigger, so you can easily detach yourself from worries or moods.

Difference between CBT and MBCT

With the help of CBT, you can identify and modify patterns of negative thoughts that often cause anxiety and depression.

On the other hand, MBCT teaches you how to identify negative thoughts and to know for a fact that these thoughts are only thoughts, and nothing more. MBCT also goes further in applying mindfulness to be aware of what is happening in the present moment, such as your current thought, your present feelings, and everything you are experiencing in the present. It helps you not to be caught off guard by any negative thinking.

CBT applies cognition to understand how negative thought works. It is often described as "a thinking therapy"; it analyzes your thoughts, emotions, and reactions. Although it takes account of the response of your body to the stress of negative thoughts, it is a therapy that deals mainly with the thinking process. The main focus of CBT is on you mentally avoiding negative thoughts.

The techniques applied in MBCT are a little different from those used in CBT; they involve things like focus on breathing, where a few minutes is spent with your attention solely on the process of your breath, and body scans, where time is spent observing the different sensations and tensions in your body during sitting meditations. Because of these techniques, it is often referred to as "a feeling process". MCBT is, therefore, both experimental and analytical; it is more focused on the body than CBT. The pivotal point of MBCT is allowing your thoughts to come and then letting them go.

Similarities between CBT and MBCT

· Some similarities between CBT and MBCT include:

· Both methods help you to manage your thoughts properly.

· They both make you more resistant to automatic thought patterns, reactions, and feelings.

· Both methods of treatment require only a short time to achieve their goals.

· They are both best suited as the only method of treatment for mild anxiety and depression, unlike treatment plans for abuse and trauma that might require more than one form of therapy and a longer time of treatment

It is important to note that both methods of treatment are more beneficial after a successful application of talk-therapy treatment. MBCT is the more helpful of the two for people who have long-term depression and need a remedy for recent episodes of depression. Even after the therapy is over, negative thoughts are still connected to negative moods in your brain and could be triggered again. So being able to monitor those triggers and your views around situations that serve as triggers is a technique MBCT provides.

Benefits of MBCT

More control over your thoughts

MBCT has helped a lot of people with various mental health issues. Currently, it is applied to teach people how they can better understand their thoughts, patterns, and mechanisms. This helps them to recognize the signs and symptoms that point towards a mental health issue.

MBCT encourages you to be mindful of the present in general, not only during the time of therapy sessions and while doing meditations. This enables you to live outside your head, to pay more attention, and connect with people around you. With this way of living, you are less likely to encounter any negative thoughts that might lead to a mental health problem. People who practice MBCT let go of any depressive thoughts instead of holding on to them.

Stress reduction

In addition to meditation, deep breathing is another practice of mindfulness that is embedded in MBCT. A deep breath is a useful technique that calms the nervous system in times of stress. This can come in handy in times when you have the urge to react to those stressors.

Stress, in general, can be reduced with the help of MBCT because it gives the ability to become more aware of yourself in the present. So, your attention is focused on the matters at hand, giving no free time to excessive thinking and worrying about certain situations in the future or past. These factors have enabled MBCT practitioners to be more resistant to stress and to deal better with any stressful situation.

Improved mood

With the joint effort of both CBT and MBCT, you can learn to improve your mood and deal with depression. Even people with anxiety and depression can learn how to apply techniques from MBCT to prevent those minor feelings of sadness from turning into a deeper state of grief.

Constant practice of mindfulness has proven to be useful in helping people connect to their purpose in life; thus, they have no time to feel worthless or lost. This is because mindfulness teaches people to be and live in the present and be more thankful for everyday life. When you pay more attention to what is going on presently rather than letting yourself be carried away with thoughts or worries and external distractions, you will not only be more thankful; you will also notice your value to the world. Some studies have shown mindfulness to be useful in developing the area of the brain that reduces anxiety and increases positive feelings

CBT and MBCT have been proven through many studies and much research to be exceptional in treating depression and anxiety, among many other mental health issues. If you are confused about which therapy method would be suited to you, ask the opinion of your therapist.

Chapter Nine: Three Mindfulness Meditation Techniques You Should Know

Mindfulness makes it easy to understand your thoughts as well as behavioral patterns. It encourages you to appreciate the little joys of life without getting bogged down by the usual stress. By being mindful, you will be more adjusted and less judgmental towards yourself, others, and any situations in life. By figuring out the connection between the downward spiral and negative thinking, you will no longer feel helpless and will be better equipped to deal with your life. It encourages you to stop harboring ridiculously high expectations from yourself while enabling you to love yourself for who you are.

From the previous chapter on mindfulness and its connection with CBT, you should have understood how mindfulness meditation is an effective way to manage our feelings of stress and anxiety. It can be used to achieve a relaxed state during panic attacks, as mindfulness helps to slow down racing thoughts while focusing on the present, letting go of negativity, and calming both your mind and body. For a better understanding of these techniques, let's start with understanding meditation itself.

Basics of Meditation

Meditation entails staying in a relaxed position and concentrating your psyche on one idea while clearing it of all others. Your concentration might be on a sound, or your breathing, counting, or on nothing at all. A desirable aspect of meditation is that the mind does not follow every new thought that comes to the surface. Meditation, being popular, definitely has different forms and styles but all still follow specific patterns as explained below:

Keeping a quiet mind

A lot is going on in our world, and it is quite hard to keep our thinking mind quiet. However, with meditation, it is possible to keep the voices down. This means you are not focused on the things in your daily dealings that put you in a state of stress, not concentrating on your life's problems. You should know that without constant practice, you will find it hard to turn off these voices inside your head.

Being in the moment

It is essential that you learn how to keep your mind focused on the present. This is possible with meditation as all forms of meditation involve focusing on the present. You being in the present consists of experiencing each moment, then letting it go, and then moving onto the next. This takes a whole lot of practice, as focusing on the moment can be difficult because of the amount of time we spend thinking about the future or contemplating the past.

It is worth noting that meditation is widely advertised as a health-boosting practice. The reasons for this are mentioned below:

Health benefits

Meditation has provided a whole lot of positive benefits, from reducing stress symptoms to enhancing immunity. It reduces episodes of depression and anxiety. It also improves concentration.

Social Benefits

It has been reported to help improve relationships and also to enhance creativity. This goes a long way to reduce cases of low self-esteem and self-judgment, which reduce individual productivity.

Having these benefits helps you to give your best in whatever situation comes your way, be it at work, school, or at home.

Cost-effective

Meditation is not one of those self-care practices that require a lot of funding; it is practically free. Your income cannot keep you from enjoying all the benefits to be had from meditation.

Productivity

Meditation only requires a few minutes (as few as five minutes!) daily to produce all its benefits.

Putting all these reasons together, it becomes easier for you to see why meditation has become a popular complement to medical practices today.

Mindfulness Meditation

Mindfulness involves focusing on the present moment rather than thinking about the future or the past. It could be focusing on a particular sensation, not for the sake of you examining the sensation but to experience it as it is. Another example is focusing on an object, not to place judgment on it but to savor the experience of the sensation you are getting from it. In other cases, you could focus on your breathing.

Certain individual components are crucial to practicing mindfulness meditation. These components include:

Focus

This is your ability to selectively place your attention or awareness on just one of the many sensations currently bombarding your mind or body, for an extended period without getting distracted. Imagine the thousands of feelings that are coming to you right now; the wind blowing against your skin, the sound of the overhead fan, the humming of the air conditioner, the pressure from the surface you are sitting on, the taste of your mouth, the rising and falling of your belly, etc. All these are demanding your attention, and it is an outstanding skill to be able to focus on just one for some time without getting distracted. This is usually hard for most people as the world we live in is overflowing with lots of things to catch our attention as we go through our activities at work and home. It has become a hard thing to sit down to read a book for just 10 minutes!

This is going to take a lot of practice but will be worth it. It is advised that you stay away from distractions like the computer, TV, radio, and your smartphone when performing mindfulness meditation.

Sensory Clarity

The next component speaks to how well you understand the information being passed from the raw data being assembled by your mind. Sometimes, what we think we know of particular situations is not so, and these misconceptions can lead to unnecessary anxiety. So, it is crucial that you calmly understand your situation before you act. This may be likened to looking through a microscope; at first, you see through lower magnifications. Later on, after a thorough understanding of the specimen at that magnification, you can change the lens to one of higher magnifications for better appreciation of the sample. This is to tell you that the more you practice, the clearer you see.

Equanimity

This is a critical ability that involves experiencing emotions and sensations without being affected by or reacting emotionally to them. This ability is a form of "decentering", which is paying attention to and accepting all thoughts coming in, but nevertheless not reacting to them.

Mindfulness Meditation Techniques

These techniques all follow specific procedures; taking notice of a particular sensation, labeling its channel of awareness, and savoring its experience without placing judgments. "Channel of awareness" is in regard to how you have become aware of the feeling being focused on. At this point, you need to understand the distinction between inner and outer awareness. These channels are seen in both inner and outer awareness, and they include:

Seeing

Outer seeing speaks to images of objects formed on the retina of the eyes while internal seeing is in regard to your imagination.

Hearing

Outer hearing occurs through our ears, while the inner awareness is playing a tune in our mind or engaged in internal monologue.

Feeling

The external feeling talks of the various stimuli you can sense within and without your body, while the inner feeling talks of your emotions such as nervousness, fear, anger, sadness, happiness, joy, etc.

That said, some analytical techniques that you should know include:

Three-Minute Breathing Space

To perform this exercise, you can either stand, sit, or lie down. Find a comfortable position to get started with this exercise. The first step is to become fully aware of yourself. Concentrate on what is going on in your mind and how you are feeling. Stop whatever activity you were engaged in and shift all your awareness back to your body, thoughts, feelings, and your breathing. Avoid moving your body and slowly concentrate on yourself.

While doing this, you might come across specific negative thoughts or beliefs present in your mind. Whenever you come across any negativity, don't try to ignore or avoid it. Instead, allow yourself to feel whatever you are feeling. Don't try to change anything at this stage. Instead, acknowledge these thoughts and allow them to pass. Now, repeat this step for any other feelings or sensations present in your body. Whenever you notice tension in a specific part of your body, acknowledge it and move on.

The second part of this exercise is to concentrate on one thing, and that is the way you breathe. Breathe in and focus on the way your abdomen moves. Whenever you inhale, your abdomen pushes upward, and when you exhale, it falls. Allow this step to anchor your thoughts and let the grounding effect wash over you. Once you have managed to gather and concentrate your thoughts and energy on yourself, you can start focusing the sense of awareness along the length of your body.

To perform this exercise, you can set a timer if you want. By setting a timer for three minutes, you will know when to start and end the exercise. The great thing about this exercise is that it can be performed anywhere at any time. Whenever you start feeling anxious, stressed, or even worried, take a break from whatever activity you are engaged in and concentrate on your breathing.

Mindfulness stretching

A great thing about the practice of mindfulness is that you can do this all day long, and it can easily be incorporated into your exercise routines as well. Before you start exercising, always concentrate on stretching your body. It helps relieve any tension or anxiety present within. Stretching is crucial because it helps reduce the risk of any injury while improving your physical performance. Apart from this, it also helps re-energize your body and prepare it for the exercise that lies ahead. Whenever you stretch, there is an increase in the supply of blood as well as oxygen to all the cells in your body. Mindfulness stretching also increases your state of awareness while bringing about balance to your physical body.

Pandiculation might sound like a complicated process, but it is a simple stretching exercise. There are three simple stages in this exercise. The first step is to pay attention to the muscles in your body while voluntarily contracting them. The second phase is to release these muscles slowly, and the third stage is relaxation. You can perform this exercise wherever you are, even when you are lying down. Try to contract all the muscles in your body, slowly release them, and then feel relaxation wash over you.

While stretching, ensure that you are stretching the right muscles; avoid placing unnecessary stress on your joints or muscles, and stretch slowly. By following these simple precautions, you can ensure you don't injure yourself while stretching or cause any pain.

There are different yoga poses you can include in your exercise of mindful stretching. In fact, most of the yoga poses include some form of stretching.

Body Scan

To start this exercise, you can either lie horizontally on the ground with your face and torso facing upward or sit on a chair. If you are sitting on a chair, ensure that your feet are planted firmly on the ground while your hands lie on your thighs. Choose a comfortable position to start the exercise.

Allow your concentration to focus solely on your body, and avoid fidgeting or moving around during this exercise. Only make deliberate movements when you have to readjust your position.

This technique primarily uses your breath to create a center of awareness. Use your breath to concentrate on your body. Don't try to change the way you breathe; once you become aware of your breathing, the next step is to pay attention to your body. Observe how you feel within your skin and observe the way your body feels. Notice how the surface you are lying on (or sitting on) feels, your surroundings, and your body temperature. While doing this, become more conscious and aware of any pain, soreness, tiredness, or tingling sensations in different parts of your body. Also, make a mental note of different parts of your body where you don't feel any sensation or are extremely sensitive.

While performing a body scan, you need to concentrate on every single part of your body, from the tips of your toes to the crown of your head. Don't ignore anything. Slowly shift your focus from one part of the body to another and take note of how you feel.

Once you have taken stock of every single part of your body, it is time to end the body scan. To do this, slowly bring your awareness back to your surroundings. Shift your focus to your breathing by concentrating on the way breath enters and leaves your body. It is time to slowly open your eyes and get back to the real world.

Daily Mindfulness

MBCT prescribes different techniques of mindfulness you can perform in your everyday life. These activities help make you more aware of your body, mind, and any emotions or feelings you experience. Once you become aware of all these things, it becomes easier to change any undesirable beliefs or emotions. Mindfulness can be practiced while showering, eating, exercising, washing dishes, even while making your bed in the morning.

As mindfulness requires a lot of practice, practicing everyday mindfulness is the best way to teach it as a lifestyle. Taking the opportunity to practice mindfulness whenever you are presented with it will help you maintain a healthy sense of awareness and balance throughout your day. This is seen in:

Mindful showering, which is about keeping your attention on just what you can see, hear, feel as you take your shower. While showering, most of us tend to think about different things. Avoid doing this. Concentrate only on the way the water feels on your body. Imagine that the water is washing away all your stress and anxieties — concentrate on cleansing your physical body, and nothing else. Pay attention to the temperature of the water, the way soap feels on your body or any other sensation you experience while lathering yourself.

Mindful eating is keeping your attention on whatever it is you are eating. Whenever you are eating, ensure that your entire concentration is on the meal you consume. Get rid of any electronics or other distractions, which will allow you to focus on the task of eating. Slowly chew your food before you swallow it. Learn to savor every morsel you eat. It is a great way to become more aware of the kind of food you feed your body.

Mindful dishwashing should only be done when you have a few dishes to wash. Mindful dishwashing is watching yourself clean the dirty dishes and listening to the sounds of dishwashing, such as water flowing. You can even pay attention to the smells if you are okay with that.

Mindfully making your bed is done by moving deliberately and with purpose while making your bed. Try to make your bed carefully and deliberately. If you usually are quite quick and careless while doing this, start paying attention to the task at hand. Even if it is a rather mundane activity, it is a great way to bring awareness to yourself. Concentrate on the texture of the sheets, the softness of the mattress, or even the way the pillows look. Put your all into what you are doing, as uninteresting as it may seem.

Pay attention to your muscle tone, your breathing patterns, and your gait. We tend to get trapped in the distractions of heavy breathing and pain during exercise; try to give yourself an experience without all these. Practicing daily mindfulness is an opportunity to maintain awareness and create balance throughout your day.

It is quite important to spend time with your loved ones, but you also need a little time for yourself. Take a break from everything and spend some time with yourself. During the "me" time, avoid any distractions; keep your phone away, don't check your emails or watch TV. There is time to get back to these tasks later. For now, concentrate on how your body feels, your thoughts, and any emotions you are experiencing. Forget about the external world and tune into yourself. It is a great way to practice self-love. Once you become aware of all this, it becomes easier to heal your body.

Mindful observation is a simple exercise that enables you to connect with everything going on in your surroundings. Most of us are often in a hurry, and we miss out on the little things in life. Start by choosing an object present in your immediate surroundings and focus only on that object for a couple of minutes. You can concentrate on a flower, tree, cloud, or anything else that you want. While doing this, carefully observe everything about the object. Once you feel calmer and no thoughts are running wild in your head, it is time to get back to your normal life.

Mindful awareness helps increase your awareness as well as build an appreciation for all the routine activities you perform. Think of any activity you perform several times daily. Perhaps it could be opening a door, drinking water, or anything else that you might have simply taken for granted. Stop for a moment and think about how you feel whenever you perform the activity. How do you feel when you are drinking water? How do you feel when you switch on your laptop? The next time you come across something that makes you smile, learn to appreciate it. It could be something as simple as sharing a meal with your loved ones or having a comfortable bed in which to sleep at night. Instead of going through your life on autopilot, take a couple of minutes to appreciate all the good in your life.

Mindful listening is the ability to listen without any judgment or bias. Our reactions, perceptions, and thoughts about most things that we see and hear daily are all based on our past experiences. Once you learn the skill of mindful listening, you can listen to everything you come across from a neutral perspective. You can start with something as simple as listening to songs. Don't judge a song based on its lyrics, genre, artist, title, or anything else. Instead, simply listen to it and allow your mind to explore the music. Allow yourself to get lost in the rhythm and sound. The idea is to let go of any preconceived notions and try to get involved in the present.

Mindful immersion helps create contentment in the present. It is about living through a routine instead of just getting things done before you move onto something else. Don't think of decluttering as a tiresome chore, and instead pay attention to all the little details that go into this activity. The aim is to try and find new emotions while performing repetitive tasks. When you become aware of all the things that you do, your willingness to do them increases while elevating your overall experience.

Mindful appreciation is quite straight forward. Take a couple of minutes daily and notice any five things you haven't appreciated in your daily life. This exercise helps you become more appreciative of all the seemingly insignificant things in your life. Most of us forget about all the little things in life because we are concentrated on attaining the goals. Learn to be grateful for every single aspect of your life. You probably have things that you wished for a couple of years ago. So, why don't you be grateful for all that you have now? Instead of being filled with regret later, it is better to be a little appreciative right now.

Avoid being judgmental. Mindfulness is your ability to accept everything about yourself. Accept the feelings, thoughts, and sensations you experience. Even all those things that you might have labeled as dangerous and self-destructive are still a part of you. Instead of ignoring them or stuffing them away in a dark corner of your mind, accept them. Merely accepting these things will not make them come true. Learn to understand that your thoughts are just thoughts. Unless you act on them, they will not become real. So, don't allow yourself to be overburdened by all this. Once you let go of all this, you will feel better about yourself. All the stress you used to experience will slowly melt away.

Regardless of the task, you are performing, ensure that all your attention is focused on the task at hand and nothing else. By doing this, you will not only be able to give 100% to the activities you are engaged in, but also improve your self-awareness. If you want to increase your efficiency as well as effectiveness, then start being mindful daily.

Other Meditative Practices

Other meditative practices generally involve two basic categories of focus: the concentrative and the non-concentrative. The concentrative speaks to having a particular object in the center, such as a candle flame, while the non-concentrative has a broader focus, such as the sounds in your environment. Note, however, that some of these focuses do have an overlapping of categories. Below is a short description of some of these practices:

Basic Meditation

Meditation involves sitting in a relaxed position purging your mind or concentrating your psyche on nothing.

Focused Meditation

This is just the primary type but you have something you are focusing on, though you are not to engage your thoughts or attention on it.

Spiritual Meditation

Though meditation is not specific to any one religion, it can be a spiritual practice. Prayer to seek guidance or inner wisdom can be a form of meditation to many people.

Things to Keep in Mind as You Meditate

- Consistent practice matters more than long inconsistent practice, but to get the best results, having a short daily practice with an occasional long exercise such as going to a mindfulness retreat is advisable.

- Regular practice matters more than a perfect method, as any meditation is better than none. So do not waste time trying to figure out the details of the technique; start! Everything else will fall into place.

- Accept that it is normal for your mind to wander even when meditating.

To conclude, don't wait any longer – get into a comfortable sitting posture and start meditating!

Chapter Ten: Don't Panic! How to Stop a Panic Attack with Mindfulness

Panic attacks are sudden, severe surges of fear, panic, and anxiety; they are overwhelming, and people with a panic attack can show both physical as well as emotional symptoms. It involves sudden feelings of terror that strike without warning, and it can occur at any time, even during sleep. Panic attacks might make you think you are dying, going crazy, or having a heart attack. However, this might not be real; the fear and terror may be unrelated to what is happening around you and not be in proportion to the actual situation.

Signs and Symptoms

Panic attacks present with symptoms such as difficulty in breathing, quivering, profuse sweating, and a throbbing pulse. In other cases, you may experience chest pain, or feel detached from yourself.

Panic attacks may occur when you are calm or anxious. Although the panic attack is a symptom of panic disorder, it is normal to have panic attacks in the context of other psychological disorders. For example, if you have a social anxiety disorder, you might have a panic attack before giving a speech at a conference. If you have an obsessive-compulsive disorder, you might have a panic attack when prevented from engaging in a ritual. Panic attacks are not pleasant and can affect social behavior.

Panic attacks are the onset of severe fear or discomfort that reaches the highest point within minutes. You can know if you're having a panic attack if you have at least four of the symptoms below:

Shortness of breath.

A feeling of being choked.

Pain in the chest region.

Unsteadiness and nausea.

A pounding heart, clear palpitation, or an accelerated heart rate.

Trembling or shaking, and sweating.

Abdominal problems.

Dizziness, feeling light-headed or faint.

Paresthesia (numbness or tingling sensations).

Feelings of being detached from reality or yourself.

Chills.

Being afraid of losing yourself.

Panic Attack Versus Panic Disorder

Having a panic attack doesn't necessarily mean that you have a panic disorder; they are quite different. One in three adults will experience at least one panic attack in their lifetime, but most of them will not have panic disorder.

A panic attack can come from being stressed. A few other diseases such as phobias or post-traumatic stress disorder can also present with the symptom of panic attacks. For example, in post-traumatic stress disorder, a panic attack can happen when a person goes back to the place the trauma occurred. These people are usually scared of their shock and not of the panic attack itself.

How to Know if you have Panic Disorder

There are several ways to help you figure out if you actually have a panic disorder and are not simply experiencing a panic attack. Some of these include:

If it happens a lot.

If you are prone to experiencing a lot of fear of having another attack.

If it usually comes on unexpectedly.

If you find yourself sitting near exits or bathrooms, so you have an easy escape route in case you get an attack.

If you are scared that certain bad things might happen if you get an attack, like being embarrassed in public.

If you avoid specific locations or situations and only allow yourself to experience them if you have a friend or family member with you or certain items like medications.

If you are avoiding physical activities, food, or day-to-day activities because you fear they might trigger a panic attack.

If you have any one or more of these, then it would be a good idea to see a doctor.

Anxiety Attack Versus Panic Attack

Most people use the terms anxiety and panic attack interchangeably, but they are two different experiences. The DSM-5 describes the features of panic disorder or panic attacks that occur due to another mental disorder. Panic attacks begin to subside after reaching their peak level of intensity at about 10 minutes. In contrast, anxiety is used to describe a core feature of multiple different anxiety disorders. The symptoms that result from being in a state of stress (such as restlessness, shortness of breath, increased heart rate, and difficulty in concentrating) may feel like an attack but are not generally as intense as those experienced at the height of a panic attack.

Who it Affects

Panic attack or disorder can affect anyone, but there are certain groups of people that it affects more often than others.

• Females: like most other anxiety disorders, mature females are more likely to experience a panic attack or disorder than grown males.

• Adults: panic disorder often appears in mid-twenties, although it can happen at any age. Most people with a panic disorder experienced the onset before the age of 33. Though it can exist in kids, it's often not noticed until they are matured.

• People who are suffering from chronic illness: most people with panic attacks or disorder report having at least one other diagnosed chronic physical or mental illness.

• You are at a higher risk of having a panic disorder if you have a family history of such.

Causes

People with specific genes are susceptible to panic disorder. However, the particular genetic patterns associated with high susceptibility have not been identified. You are at a higher risk of developing panic attacks if either or both of your parents have been diagnosed with depression, anxiety, or bipolar disorder.

Panic attacks can be triggered by:

• Work stress

• Social stress

• Various phobias

• Withdrawal from drugs or alcohol

• Chronic conditions or pain

• Medications or supplements

• Driving

• Caffeine

• Memories of severe trauma that happened in the past

Duration of a Panic Attack

Although the time varies between individuals, panic attacks typically reach their highest point within ten minutes or more, and then symptoms begin to decline. Panic attacks seldom last for more than an hour, with most lasting for around thirty minutes.

How often does a panic attack happen?

It is different for different people, you might have one panic attack and never experience another, and you might have attacks once a month or even several times a week.

Can a panic attack kill you?

Panic attacks cause different issues, and many people feel they are about to die when they experience it. However, having a panic attack cannot kill you.

Ways to Stop a Panic Attack

Mindfulness is related closely to meditation and can be practiced at any time, whether you are walking, taking a rest, or working out. Mindfulness is like meditation in motion. Mindful people are optimistic about the present, and they keep an open mind. They are not contemplating or giving a thought to things of the past, nor are they worried about what the future holds. Mindfulness requires that you keep a mind without worries. You will need to gather your focus and perception from inside your head to outside your head because there are a lot of more exciting things on the outside. You can practice mindfulness while walking and working outdoors and during sports. It is mindfulness that will help you shift your attention away from the pain endured during exercise to having a pleasant sensation. Mindfulness will change your perspective of whatever situation you apply it to, and consistent practice of mindfulness will eventually improve your thought patterns and your general mindset.

Here are a few steps to stop a panic attack:

• Breathe deeply.

• Recognize it as a panic attack.

• Close your eyes.

- Practice mindfulness.

- Focus on an object.

- Relax your muscles.

- Find your happy place.

- Engage in light exercise.

- Repeat your mantra.

- Take benzodiazepines.

Recognize that it's a panic attack and not a heart attack

Panic attacks come with the symptom of thinking there is a danger ahead of you or that you are dying. These symptoms can be scary, but the first thing to do is to take away this fear and acknowledge that you are merely having a panic attack. Ascertain that there is no impending doom and remind yourself that this is temporary, it will pass, and you will be okay. This acceptance will allow you to focus on other techniques to treat your symptoms.

Deep breathing

The first practical way to deal with a panic attack is to practice deep breathing. Focus on breathing in deeply and slowly through the nose until the air fills your chest, then sigh out through your mouth. After every few breaths, you should relax to regain your rhythm. The essence of deep breathing is controlled breathing. It is vital you take note of the counts for optimum breathing and to prevent hyperventilation. Hyperventilation can worsen other symptoms; however, if you can control your breath with the counts, you are less likely to experience this.

Shut out any visual stimuli

A loud environment, with various visual stimuli, can trigger your panic attack. Once you feel you have a panic attack, closing your eyes will reduce these stimuli and prevent taking in more visual information so you can easily focus on breathing and controlling the attack.

Focus object

Focusing on a single object can be beneficial during a panic attack. Look for an object that is close to you and carefully analyze it. Trying to explain it will shift your mind to the object and take your mind off other symptoms of the panic attack. For instance, you may choose to analyze how a shoe is placed in its rack carefully. If it's improperly placed or doesn't fit in, you can try describing how it should be to yourself, or what kind of shoe would fit in its place. With this quick focus, whatever panic symptoms you are feeling may subside.

Mindfulness

A panic attack can quickly get you detached from reality. Keeping your mind in the present can redirect your senses away from the intensity of the anxiety. You can give yourself a little task. Identify four things that are around you, feel the texture of three objects, listen to two different sounds, or smell something that can trigger a memory. This exercise aims to keep you grounded in reality and not moving from one worry to another.

Muscle relaxation techniques

A popular technique for coping with panic attacks is muscle relaxation. Sometimes your muscles may become tense unconsciously in response to what you feel to be a dangerous situation. Muscle relaxation techniques help you to control your body's response. In this technique, you repeatedly flex your muscles and then relax them. After the relaxation, you should remain seated to allow yourself to become alert again. This technique is most effective when you have practiced it before a panic attack occurs.

Happy place

You probably have a place or a view that makes you feel utterly relaxed. When you have a panic attack, you can close your eyes and picture yourself in your happy place. You can create a mental picture of the beautiful view that got to you when you were on the plane, or the tranquility you get any time you are listening to music at the beach. However, try not to think of noisy areas like a busy park or a crowded street, even if these are places you enjoy in your everyday life.

The internal mantra

Ever seen someone about to have a panic attack just close their eyes and start moving their lips? Most likely, they're repeating a phrase or two in their head that helps them deal with the attack. Repetition of a mantra, even without speaking it aloud, can be relaxing, and it gives you something to hold on to during a panic attack. Find one that clicks with you and that you can easily remember; repeat it continuously until you feel the panic attack starting to dissipate.

Light exercise

Endorphins are a lifesaver. When our brain releases this chemical into our blood stream we feel happy and energized. Light exercise does wonders when it comes to flooding our system with endorphins, and this ultimately improves our mood drastically. You can choose a light workout that is gentle on the body when you are stressed, like taking a stroll or maybe a quick jog around the park.

Benzodiazepines

Although benzodiazepines can be addictive, they are a medication that may help you treat panic attacks. Just remember that the body can easily adjust to it over time, and it should therefore only be used on rare occasions and in cases of pressing need. When you are having serious panic attacks and can feel one coming on in the worst of situations, that's when this will come in handy.

Keep a diary

Keeping a note of what happens every time you get anxious or have a panic attack can help you spot patterns in what triggers these experiences for you, and this will, in turn, help you to think about how to deal with these situations in the future. You can also try keeping a note of times when you can manage your anxiety successfully; this can, in turn, help you feel more in control of the stress you feel.

How to Help Someone with a Panic Attack

If you ever encounter a person suffering from a panic attack, do not fear. You should not add to the person's stress by panicking or shouting at them. Instead, you can try some of these ways to help.

- Stay calm, do not scream or add your fear to the person's distress.

- Ask questions. If it's not their first time, ask if they use certain medications and if you can help them with it.

- Don't think you know everything that is going on. Ask them the cause of the panic and what they need.

- Be encouraging and positive while talking to them in simple sentences they can easily understand.

- Prevent the person from going into hyperventilation by encouraging them to breathe more slowly and deeply.

You can also say to the person things like:

- "What can I do to help you get through this?"

- "You are doing fine, and I am proud of you."

- "This attack is not dangerous at all, although it might feel scary."

Take this simple approach, and you will find out that you can:

- Reduce the amount of stress in a very stressful situation.

- Prevent the worse from happening in the situation.

- Curb a complicated experience.

You can help someone recovering from a panic attack by:

- Giving the person the autonomy to proceed in therapy at his or her own pace.

- Being patient and addressing all efforts toward recovery, though the person may not meet all of the goals.

- Avoiding things or situations that can cause anxiety.

- Not panicking, even when the other person panics.

It is all right to be concerned and anxious yourself, but you can control yourself and the situation.

Fortunately, panic disorder and panic attack is a treatable condition, even to the extent of complete disappearance. Psychotherapy and medications have been used as effective treatments, either singly or combined. If another medication is necessary, your doctor may prescribe medicines for anxiety. There are certain antidepressants or anticonvulsant drugs that also have anti-anxiety properties, and a type of heart medication known as beta-blockers, which help to prevent and control the episodes of panic disorder.

Chapter Eleven: How to Prevent a Relapse

Lapse: A lapse is a brief return to feeling down or to your old habits. It is a common and temporary situation.

Relapse: As opposed to a lapse, a relapse is a complete deterioration or complete return to your initial state of health after a temporary improvement.

For example, you had a phobia of spiders, and now you know that it is best not to scream when seeing one. Somewhat, you calm yourself down, breathe, tell yourself some coping thoughts, and gradually ignore the spider. So, if you find a spider in your room one day and you scream, that is a lapse. If you then go back to screaming and running whenever you see a spider, then we can call that a relapse.

Lapses can progress to relapses, but this should not necessarily happen. You can stop a lapse from escalating into a relapse.

When Does a Lapse Become a Relapse?

The general belief that what you say to yourself after a failure can make or break you is very much applicable here. What you think and say to yourself after a lapse can lead you back to the right track or throw you into relapse. Seeing a lapse as a failure can keep you sick and lead to a relapse. A better perspective is that you were able to have emotional wellness before, you can have it again; process whatever happened before and learn from your mistake.

Going back to our spider-phobia example: if, after avoiding the spider all day, you said to yourself, "It looks like I'm bringing back old habits; I need to do better tomorrow and get myself together!" you would discover that your lapse would probably decline or stop completely, and now you can face your anxieties and fears head on. If you avoided spiders all day, and at the end of the day said to yourself, "All my hard work is a waste, now I'm here again. Arggghhh, I'm such a jerk! Why am I even trying when there is no cure?" This is not really helpful, and it won't help your recovery.

Can I Prevent Lapses and Relapses?

Yes, you can prevent lapses and relapses, and here are seven clues you can use:

Do not give up on practicing; the best way to prevent a lapse is regularly practicing your CBT skills. If you are practice regularly, you will be in good shape to handle whatever situations you might face.

Understand yourself (Red Flags). Relapse doesn't happen suddenly. It occurs over a period of time. Preventing relapse by understanding yourself is not complicated. Understand yourself by identifying your triggers, asking for help, and sharing your feelings.

New challenges. Everyone is a work in progress, and you are no exception. This means there is always a chance to get better, and you can work on yourself and live a more fulfilling life. It will be less easy to backslide into your old ways if you deliberately work on new ways of overcoming your anxiety. An excellent way to prevent lapses is by challenging yourself regularly and taking up new scary situations. Make a list of cases that sound scary to you and initiate anxiety when you think of them, and work on them.

Learn from Your Past Experiences. Lapses are not synonymous with failure, rather they are opportunities to learn and get better. Figure out the situation that always leads to you having a lapse and make a plan that will help you deal with these situations better in the future.

As said earlier, what you say to yourself after a lapse can impact your behavior. Have a few positive things that you say to yourself. CBT has helped you, and you cannot throw away everything you have learned. Going back to the beginning means having anxiety and not knowing how to handle it. Going back to practicing your CBT skills will help you to master your anxiety again in a short time.

Be kind to yourself; remember that lapses are not the end of your world, take it easy on yourself and learn. No one is above making mistakes; we all make mistakes. We all try to speak nicely to people, so do the same to yourself; don't say harsh things to yourself. Lapses can be a blessing in disguise at times because you get a chance to learn that you can go back to fashion out a new formula of dealing with your situation.

Enjoy yourself; make sure you always take the time to rest and relax from all the hard work you are doing. Appreciate yourself; buy yourself a nice meal, get something new, or hang out with your friends. You can also reward yourself by pampering yourself and taking some time to relax.

Depression Triggers

People who have a history of depression can have triggers that cause a depressive episode. However, an event being stressful for a person does not mean it will trigger depression. Triggers vary from one person to another; what is difficult and stressful for you might be easy for others.

Potential Depression Triggers Include:

Sad occurrences

Various life situations, such as the death of a loved one or a tragic end to a treasured relationship, can be a trigger for depression. According to a study, 20% of people enter into depression after this type of loss.

Stopping Treatment

Most people stray away from treatment after they feel their symptoms are getting better. A high percentage of these people gradually see their symptoms setting in again, and they may enter into another episode of depression. Finishing your treatment can surprisingly lower your risk of relapse.

Traumatic events

Remembering events that have caused trauma in the past can bring about a relapse. People who have had depression resulting from attacks or disasters are at a high risk of entering another episode.

Health Conditions

People who have been diagnosed with particular health conditions can lose their self-esteem and confidence. They may enter into overthinking and, consequently, into depression. If you find yourself in this category, take care of your medical condition and prevent it from taking over your life. This will give you control over the depression.

Financial issues

Money problems are prevalent causes of worry. A way to avoid this is to practice a healthy economic lifestyle. Create a budget and stay true to the budget. Also, you might want to create a savings plan, so you are not tempted to spend all your money at once. Attend programs that don't cost a fortune so that you can spend time with family and friends. Increased financial stability can reduce your risk of having a relapse.

Other factors you need to identify and avoid include:

Hormone changes

Addictive behaviors

Sexual problems

Poor sleep habits and diets

Feeling stressed and overwhelmed

Ways to Minimize Depression Triggers

Not all depression triggers are inescapable; some can be avoided.

It is best you learn how to find your way around these triggers as much as you can. If you are starting to get overwhelmed, here are some steps that can help you out:

Stay Positive

Find ways to improve your self-esteem, and regularly say encouraging words to yourself.

Be social

Relate with friends, family, and reach out to them when you begin to feel your symptoms getting overwhelming.

Avoid Alcohol

There is a false belief that alcohol makes you feel better; though it may seem like it, the truth is that it can make your depression worse.

Early Signs of a Depression Relapse

If you've had a history of depression, symptoms might start appearing again and trigger worry; this is totally understandable. People who have experienced depression before may have a recurrence after a period of time. This period can range from weeks to years, sometimes many years after the occurrence. If you can spot the red flags early, you might be able to stop severe episodes from occurring.

About half of the people who overcome an episode of depression for the first time will remain well. For others, depression can come back a few times throughout their lives. People have different degrees of recurrence; some experience sadness or just want to avoid daily activities. However, if you have these feelings almost daily for more than two weeks, and it begins to affect work or social life, then you may be experiencing depression.

Two ways depression can return are:

When symptoms start to appear again or worsen during recovery from an earlier episode, we can say relapse is looming. Relapse is likely to occur within two months of stopping treatment.

Most recurrence occurs within the first six months after recovery from the previous episodes.

Around 20% of people usually experience a recurrence, but this can rise when depression is severe.

We have some depression-like disorders that can return frequently. These include:

Seasonal affective disorder (SAD): SAD occurs mostly during the winter months, due to the decrease in sunlight.

Premenstrual dysphoria syndrome (PDS): PDS is a severe form of premenstrual syndrome.

Early Signs of Depression Relapse

Some people have their depression symptoms occur once; for others, it can occur over and over. It is essential to pay attention to your symptoms when they occur because this will quickly help you catch a possible sign of relapse. Early signs that you might have a relapse include;

Having extremes of sleeping disorders; excessive sleep or lack of sleep

Loss of interest in activities you enjoyed doing before

A depressed feeling of sadness and anxiousness

Memory issues

Feeling regrets over past events

Thoughts of or attempts at suicide

Avoiding social conversations and relationships with people

Extremes of appetite leading to excessive weight gain or loss

Suicide Prevention

People who commit suicide must have talked about it once or more in their conversations, no matter how serious it did or did not sound. Do not ignore these signs. Many of these people try to seek help, and they want the pain to stop. Take any suicidal talk seriously and try to yield to their cry for help. Here are a few tips to respond to someone if you notice any signs of suicide.

Listen to their conversations honestly, and if you're unsure if they are suicidal, ask nicely.

Respond quickly to severe suicidal risk. Put a call through a local crisis center, call 911, remove harmful objects from the area, and do not leave them alone.

Get professional care and follow-up treatment.

If you are having suicidal thoughts, don't use suicidal talks to give someone an idea that you are thinking about suicide. Instead, open up honestly, and you can save a life.

Tips for Preventing a Relapse

People suffering from episodes of depression can have crushing, intense feelings. The following strategies can help prevent depression relapse:

Have supporting relationships.

Avoid isolation. It is imperative to surround yourself with understanding, kind, and supportive people.

Avoid and modify depressive thinking patterns.

CBT can help you change your thinking style. Most people suffering from depression have negative thinking patterns. These patterns can be changed, and CBT techniques we've discussed can be useful for you for a lifetime.

Follow your prescribed medication.

Work together with your psychiatrist and follow any treatment pattern they give you.

Be ready for a relapse. It is advisable to plan for relapse and act upon signs as quickly as they appear.

Correcting and Coping with a Relapse

Having a return to unhelpful anxiety reactions and old thought patterns might mean that the initial treatment is not working effectively. We recommend that you seek your doctor's advice and preferably change your treatment strategy. Another treatment option is the use of medications like antidepressants or mood stabilizers with the doctor's prescription. If you've been on medication before, and it seems to be failing, you can talk to your doctor about a change in dosage.

Conclusion

The effectiveness of CBT is one of the primary reasons why it is being used all over the world to improve mental health. Long gone are the days when doctors and therapists used to focus solely on drugs and other pharmaceuticals as treatment. CBT can help treat anxiety, personality disorders, depression, and other behavioral problems associated with mental health.

Prioritizing your mental health is as important as taking care of your physical health. A healthy body will not do you much good if your mind is continuously plagued with negativity. You can always seek help and try cognitive behavioral therapy to improve your mental wellbeing. Now that you are armed with all the information you need, it is time to get started as soon as possible!

Thank you and all the best!

References

Ben, M. (2019). In-Depth: Cognitive-behavioral therapy. Retrieved from

https://psychcentral.com/lib/in-depth-cognitive-behavioral-therapy

Rosy B, et al. What to expect in CBT. Retrieved from
http://cogbtherapy.com/what-happens-in-cbt

Mental illness. Retrieved from https://www.mayoclinic.org/diseases-conditions/mental-illness/symptoms-causes/syc-20374968

Diagnostic and Statistical Manual of Mental Disorders DSM-5. 5th ed. Arlington, Va.: American Psychiatric Association; 2013. https://dsm.psychiatryonline.org. Accessed November 2019.

Healthline Editorial team, and medically reviewed by Timothy, J. (2017). Signs of Depression. Retrieved from https://www.healthline.com/health/depression/recognizing-symptoms

Erica, J. (2018). 11 signs and symptoms of anxiety disorders. Retrieved from https://www.healthline.com/nutrition/anxiety-disorder-symptoms#section2

Deanna, R. (2016). How to Create Achievable Goals for Your Mental Wellness. Retrieved from https://www.goodtherapy.org/blog/how-to-create-achievable-goals-for-your-mental-wellness-0822164

Mark, T. (2016). 3 Instantly Calming CBT Techniques for Anxiety: Cognitive-behavioral tools that anyone can use. Retrieved from https://www.unk.com/blog/3-instantly-calming-cbt-techniques-for-anxiety/

Chris, C. Treating depression with Cognitive Behavioral Therapy. Retrieved from https://journeypureriver.com/treating-depression-cognitive-behavioral-therapy/

3 CBT tips to help overcome workplace stress. Retrieved from https://www.efficacy.org.uk/blog/corporate-wellbeing/3-cbt-tips-to-help-overcome-workplace-stress/

Sheri, J. (2014). CBT vs. MBCT – What is the Difference? Retrieved from https://www.harleytherapy.co.uk/counselling/cbt-mbct-difference.htm

Courtney, E. (2019). What are Intrusive Thoughts in OCD & how to get rid of them? Retrieved from https://positivepsychology.com/intrusive-thoughts/

Kimberly, H, and medically reviewed by Timothy J.L (2019). Intrusive Thoughts: Why We Have Them and How to Stop Them. Retrieved from https://www.healthline.com/health/mental-health/intrusive-thoughts#causes

Mindfulness Animated in 3 minutes. Retrieved from AnimateEducate. https://www.youtube.com/watch?v=mjtfyuTTQFY

Katharina, S, and medically reviewed by Steven, G. (2019). Use Mindfulness Meditation to Ease Anxiety. Retrieved from https://www.verywellmind.com/mindfulness-meditation-exercise-for-anxiety-2584081

Ana, G et al. (2018). 11 ways to stop a Panic Attack. Retrieved from https://www.healthline.com/health/how-to-stop-a-panic-attack#recognize-panic-attack

Panic Free TV. Meditation for panic attacks: does mindfulness work? Retrieved from https://www.youtube.com/watch?v=_EbqcVH9eVg

Symptoms of a Panic Attack. Retrieved from Anxiety and Depression Association of America. https://adaa.org/understanding-anxiety/panic-disorder-agoraphobia/symptoms

How to Prevent a Relapse. Retrieved from Anxiety Canada. https://anxietycanada.com/sites/default/files/RelapsePrevention.pdf

Regina, B.W. (2016). 7 Factors That Can Trigger a Depression Relapse. Retrieved from https://www.everydayhealth.com/hs/major-depression-health-well-being/factors-can-trigger-depression-relapse/

Timothy, J. (2019). What are the early signs of a depression relapse? Retrieved from https://www.medicalnewstoday.com/articles/320269.php

Check out another book by Heath Metzger

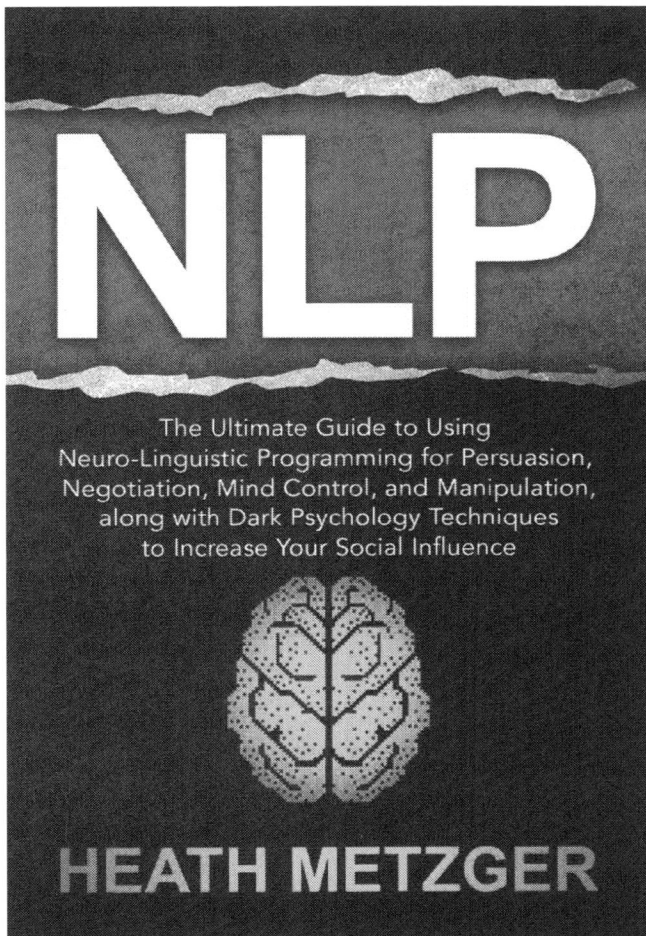

Manufactured by Amazon.ca
Bolton, ON